BECOMING RACHEL CARSON

The Origin and Legacy of a Legend

Also by Robert K. Musil

Rachel Carson and Her Sisters:
Extraordinary Women Who Have Shaped America's Environment

Washington In Spring:
A Nature Journal for a Changing Capital

Hope for a Heated Planet:
How Americans Are Fighting Global Warming and Building a Better Future

BECOMING RACHEL CARSON

The Origin and Legacy of a Legend

ROBERT K. MUSIL

Author of *Rachel Carson and Her Sisters*

A RACHEL CARSON COUNCIL BOOK

A RACHEL CARSON COUNCIL BOOK

For more information on the Rachel Carson Council
8600 Irvington Avenue
Bethesda, MD 20817
(571) 262-9148
rachelcarsoncouncil.org

ISBN 979-8-234-08067-7

Library of Congress Control Number: 2026911991

Published by:
Rachel Carson Council, Inc.
8600 Irvington Avenue
Bethesda, MD 20817
571-262-9148
rachelcarsoncouncil.org

To the Rachel Carson Council staff and RCC Fellows
who are brilliantly and boldly extending Rachel's legacy

RACHEL CARSON IS A LEGEND. She is enshrined in historic homesteads and houses, national wildlife refuges, colleges and universities, postage stamps, children's books, even a bridge. She is pictured in two Halls of Fame and her four major books continue to sell, line the shelves of bookstores, and are taught in classrooms around the world.[1]

How this small, shy woman from a poor farm in Pennsylvania rose to fame, and, for her detractors, to notoriety, has intrigued, even obsessed, academics, experts, biographers, media producers, and environmentalists for decades. How did a girl who roamed the fields and woods near her home, counted birds and animals as her friends, and read aloud to her dog, become Rachel Carson — a name synonymous with the best of boldness, bravery and advocacy on behalf of people and the planet? What are the sources of her lyrical and luminous prose, her imagination and empathy, her cutting edge science, her fearless stands on controversial issues, even her profound

influence on policy during the Kennedy years and after? Throughout her career, how did Rachel Carson manage to work in the face of rejection and ceaseless personal and political attacks from critics and corporations? And, she wrote, testified, travelled, and knit together a new environmental movement — one concerned with human beings and our health, well-being, and justice, as well as the peregrines, pelicans, osprey, and eagles who share and depend upon the same endangered ecosystems as *homo sapiens*. Even long after her death, Carson not only survived, she became even more admired, more acclaimed, despite concerted conservative campaigns to label her a scientific fraud responsible for the death of millions when DDT was banned. And, importantly, what is Rachel Carson's continuing legacy in a world that is aflame with conflict and climate change? How does she continue to inspire new generations of environmental leaders, writers, scientists, academics, activists and advocates from Al Gore to Greta Thunberg?

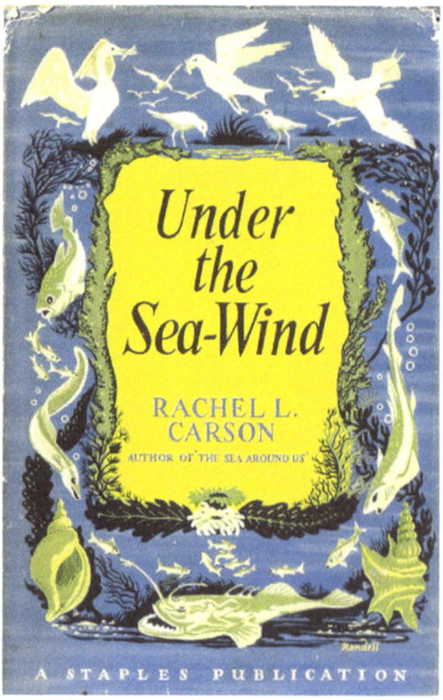

To appreciate Rachel Carson's enduring contribution to the environment, to health and justice, to literature, politics, and the world, perhaps the most revealing way to start is to take a close look at her and her work when, as a young professional in 1941, she burst, seemingly fully-formed, onto the literary scene with wild acclaim for her first and favorite book, *Under the Sea-Wind*. Published just before Pearl Harbor as the nation mobilized for total war, its sales sagged and Carson went out of print. It would be a full decade before Rachel Carson emerged in 1951 as a National Book Award-winner with a blockbuster best-seller, *The Sea Around Us*.[2] It made her a beloved household name, gave her financial independence, and allowed her to turn to full-time writing with a reputation for scientific accuracy and poetic prose that would later allow her to produce a book about poison. It was called *Silent Spring*.[3]

But it is *Under the Sea-Wind*, with its lyrical, luminous prose, its love for the mysterious and powerful sea, the struggle for survival of the creatures in and around it, its profound respect and empathy for other beings — all backed by meticulous research and science — that reveals the essential Rachel Carson and her central ethos. It is the often overlooked foundation stone of the Rachel Carson canon. *Under the Sea-Wind* takes us into an unknown, often dangerous world where Carson's careful observation, curiosity, and rich imagination, evident from childhood, draws us in. We find ourselves on an exhausting and epic migration to the Arctic with Silverbar and Blackfoot, two small sandpipers called sanderlings, and rooting for Scomber, the mackerel, feeling and fearful, as he is trapped in the horrific nets of a fishing trawler. And, incredibly, we journey with Anguilla, the eel, starting as a small elver who has survived ten years in the backwaters of North Carolina, to the Sargasso Sea that had puzzled explorers, mariners, and geographers long before it had becalmed Christopher Columbus and the *Niña*, *Pinta*, and *Santa Maria*.

Carson pulls off this feat with immense imagination, nearly Dickensian turns of plot, and vivid, pictorial prose that offers up our characters in Technicolor and sound when they had, for generations, been portrayed in books and drawings as mostly colorless, static, and mute.[4] Throughout all this, we do not simply sympathize with birds and mackerels and eels. We enter into their lives in a world where humans are merely lurking dangerously in the shadows.

If *Under the Sea-Wind*, one small volume on the animals of sea and shore, were all we ever had of Rachel Carson, it would place her in the front ranks of the great nature writers, with those, like Thoreau, still worth reading in the classroom or in a comfortable chair. It reflects Carson's upbringing, her precocious reading and writing, her keen observation and powerful imagination, and her values, especially empathy — the ability to enter into and appreciate the lives of others.[5] *Under the Sea-Wind* shows a young writer with immense talent in a fresh, unrestrained debut; Rachel Carson is in love with nature, the sea, and the mystery and meaning of it all. She has found "something to write about,"[6] seamlessly fusing her passion for poetry and literature with the excitement she finds in the natural world. "If there is poetry in my book about the sea, it is not because I deliberately put it there, but because no one could write truthfully about the sea and leave out the poetry."[7]

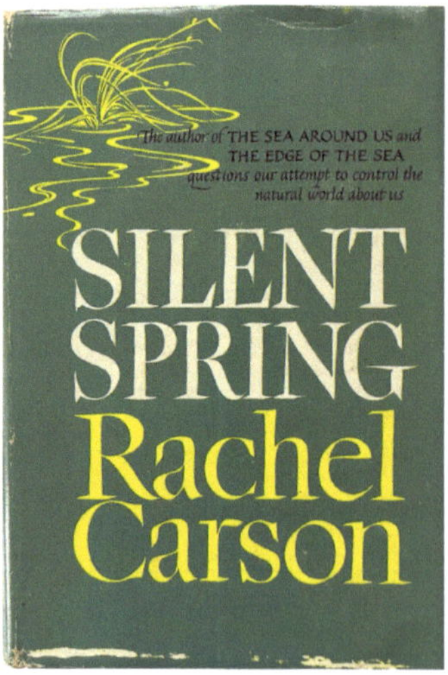

Most readers, students, scholars, and scientists know Rachel Carson as the brave and embattled author of *Silent Spring*. As many have said, it is a book that changed the world. But, in fact, it was Carson's entire life and career, her accumulated network of friends, colleagues, scientists, and environmental and conservation organizations, along with her status as the beloved writer of best-selling ocean books that allowed her to create the mastery and sensation of *Silent Spring*. Rachel Carson is central to modern American literature, history, ecology, and environmental science, as well as today's huge environmental movement. If *Silent Spring* is the pinnacle of her career, *Under the Sea-Wind* is the Rosetta stone to understanding how Rachel Carson, who signed it Rachel L. Carson, eventually became a legend.

The Birth of a Writer

Under the Sea-Wind was published when Rachel Carson was thirty-four years old, a young, but seasoned professional who began early drafts while working at the U.S. Bureau of Fisheries.[8] Carson had won literary prizes in her youth and begun college at the Pennsylvania College for Women (PCW) as an English major who was encouraged by her professor and friend, Grace Croft. Rachel wrote for the college paper and won high praise for her prose style before switching her major to biology.[9] By the time Carson initially hand-wrote and then typed out the beginning of *Under the Sea-Wind*, she was an accomplished young writer and scientist.[10] She had spent years at the Johns Hopkins graduate school in zoology and at the Woods Hole Biological Laboratory on Cape Cod. She had studied with some of the world's leading biologists, published articles on the fauna of the Chesapeake Bay for the *Baltimore Sun*, and taught biology at both Hopkins and

the University of Maryland.[11] Only the need to financially support her parents and extended family as the Depression deepened caused Carson to leave her Ph.D. unfinished. With few academic jobs open at the time and even fewer for women in science, Carson took the advice of her Chatham biology professor, friend, and mentor, Mary Scott Skinker, who had convinced her to switch her major to biology, to apply to the Federal government. She was hired by Elmer Higgins, a respected ichthyologist; he recognized her talent immediately.[12]

By 1937, Carson was promoted by Higgins to assistant aquatic biologist at the Bureau of Fisheries and assigned as a working scientist to write about research at the Bureau's labs, starting with the government fisheries laboratory on the North Carolina Outer Banks in Beaufort, then a distant oceanfront town. Carson was ecstatic at the opportunity, taking extensive notes on the lab's research, while roaming and exploring the seaside, often at night, where she would lie on the beach and listen to the sounds of the shore and survey its shadowy shapes.[13]

In an anecdote Carson loved to recount, Elmer Higgins rejected her first attempt at an introduction to studies of marine life done for the Bureau of Fisheries. It was called "A World of Waters." He returned it saying, with a twinkle in his eye, "...this will never do. It's too good. You should send it to the *Atlantic*."[14] Carson delayed and delayed, polished her text some more, and finally sent it off. The *Atlantic*, too, was quick to recognize Carson's literary and scientific gifts. They published her essay in the September, 1937 issue of the *Atlantic* as "Undersea."[15] From there, it caught the attention of the senior editor at Simon and Shuster, Quincy Howe, and Hendrik Van Loon, a popular author and illustrator, whose letter to Carson praising her work came in an envelope on which he drew waves, whales, and sharks. Carson accepted Van Loon's invitation to visit in Connecticut where he said he would introduce her directly to Howe and others at Simon & Shuster. The result was an offer to do the book that became *Under the Sea-Wind.* As Van Loon explained to Carson, he had always hoped for a writer who could bring to life what lay beneath the surface of the sea. In his frequent trips across the ocean he had been "so impressed by its apparent lifeless aspect – 'not a snout or a spout did I see."[16]

The first section of *Under the Sea-Wind*, "The Gull's Way," reflects Carson's time at Beaufort, her immersion in the life of the

shore, and her long interest in and knowledge of birds and their behavior. The opening paragraphs allow Carson to literally assume a bird's eye view of the beach. They reflect, too, the wonder and mystery of her evening explorations. A large bird emerges, as if from a painting, out of the dusky, cloudy, moonlit horizon. The sentences are sensuous, sonorous, and save the most potent phrases for last. They seem created, as Carson had done when writing them, to be read aloud. "The Island lay in shadows only a little deeper than those that were swiftly stealing across the sound from the east... Both water and sand were the color of steel overlaid with the sheen of silver, so that it was hard to say where the water ended and land began...With the dusk a strange bird came to the island from its nesting grounds on the outer banks. Its wings were pure black, and from tip to tip their spread was more than the length of a man's arm. It flew steadily and without haste across the sound, its progress as measured and meaningful as that of the shadows which little by little were dulling the bright water path. The bird was called Rynchops, the black skimmer."[17]

Critical Acclaim for Carson's Debut

When it was published by Simon & Shuster in November 1941, *Under the Sea-Wind* was instantly recognized as a special contribution to our literature. Paul H. Oehser, writing in the *Washington Post*, noted enthusiastically that "only that rarest combination among men [sic], a scientist and a poet, could have written this book." Oehser was especially taken by *Under the Sea-Wind*'s perspective on "time and tide" and prose that "moves like the sea itself, like quiet waves lapping the shore, unostentatious but fresh and basically lyric."[18] William Beebe, the leading ocean explorer and ornithologist of the time, said in *The Saturday Review* that Carson's first book makes "the sea and its life a vivid reality...Miss Carson's science cannot be questioned; I have been unable to detect a single error."[19] But Pearl Harbor was strafed and bombed a few weeks later. The lavish praise for a gifted new writer was soon forgotten as the nation mobilized for global war. Beebe, however, who over the years became Carson's friend, had been so deeply impressed by *Under the Sea-Wind* that he included its chapter on eel migration in his *Book of Naturalists*, a collection of pieces from the greatest nature writers of all time — from Aristotle and Pliny to John Muir, John Burroughs, Julian Huxley and Teddy Roosevelt. Rachel Carson was the only woman included.[20]

During the war and after, despite its critical success, *Under the Sea-Wind* sank beneath the waves, selling less than 2,000 copies.[21] It would be lost for over a decade. Rachel Carson took to free-lancing articles for magazines and newspapers as she worked at the Bureau of Fisheries that soon became the United States Fish & Wildlife Service (USFWS). Nevertheless, *Under the Sea-Wind* shows that by 1941 the young Rachel Carson was already prepared to become the beloved, best-selling author of *The Sea Around Us* (1951) and *The Edge of the Sea* (1955).[22] At the initiation of her career, Carson's debut foreshadows her ability to elicit with *Silent Spring* a powerful emotional response to the killing of birds, fish, and other wildlife with pesticides and herbicides. Others had written about the dangers of pesticides to animals and human health.[23] But only a master of both narrative and natural science, infused with the emotion, imagination, awe, and wonder, so easily visible in *Under the Sea-Wind*, could have had such a profound impact.

Carson Brings the Sea Alive

Rachel Carson was researching and writing *Under the Sea-Wind* in the 1930s at a time when satellites and computers, underwater filming, geospatial mapping, and new technologies developed during World War II had yet to be invented. Contemporary oceanography truly began after the war with explorations like Roger Revelle's Pacific expeditions in 1950 and 1952 and discoveries about the deep, dynamic ocean floor, mountain ranges and rifts, and the realities of plate tectonics.[24] Carson reviewed, read, edited, and grasped the implications of these scientific leaps, making their meaning accessible to the public in her breakthrough 1951 best-seller, *The Sea Around Us*. Yet earlier, when Rachel Carson began her research for *Under the Sea-Wind*, the ocean had long been seen as a mostly lifeless place to be crossed as quickly as possible, to ship (literally) resources and goods, and, increasingly, a convenient place to dump the burgeoning mounds of trash and waste accumulated on land.

Carson, William Beebe and Modern Oceanography

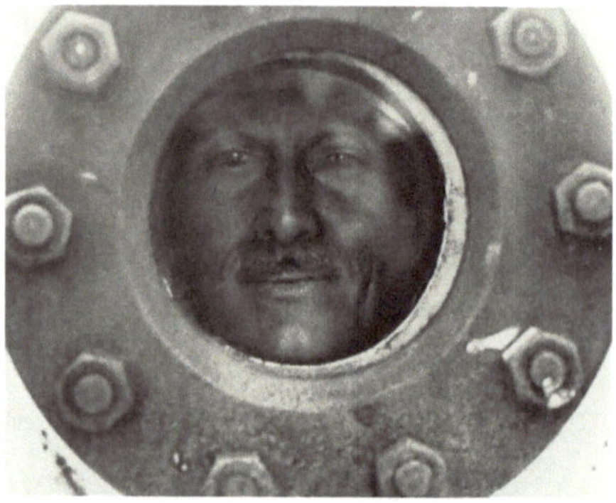

From her first excited exposure in 1929 to the resources and library of the Woods Hole Biological Laboratory in Cape Cod, Rachel Carson read widely in marine science and oceanography. From the mid-nineteenth to the early twentieth centuries, explorers and oceanographers had begun to probe and plumb the seas, map some of its depths, and discover a growing number of species. But much of the knowledge of the ocean was gained with comparatively simple instruments and soundings, including those of the Challenger expedition, the most complete and first series of investigations to circumnavigate the globe in the refitted *H.M.S. Challenger* (1872-1876).[25] Scientific knowledge of the ocean was confined to an elite, small, and specialized group of men; little was done to reach the general public. Only in the late 1920s, did William Beebe, who would review and praise *Under the Sea-Wind*, begin to popularize ocean exploration with his dives to depths of 6 to 60 feet in a homemade helmet. Then, teaming up with the wealthy engineer, Otis Barton, who helped design and construct Beebe's idea for a submersible to explore the ocean more deeply, the two innovators began to carry out test dives in what they christened a Bathysphere. On Wednesday, August 15, 1934, Beebe and Barton made themselves world famous by descending in the Bathysphere to 3,028 feet beneath the ocean surface. Beebe's daring descent was broadcast by NBC and the BBC, described in popular magazines, and captured more fully, with photographs and eight plates in color, in his 1934 book *Half-Mile Down.*[26] Beebe, who inspired both Rachel Carson and Sylvia Earle, the leading ocean explorer and marine scientist of our time, made a number of scientific discoveries and stirred public interest in the ocean. But his dramatic style and his ability to promote his work made some scientists skeptical of his findings. And, despite his depth records, Beebe had restricted views, "... my forehead pressed close to the cold glass,-that transparent bit of mother Earth which so sturdily held back nine tons of water from my face." He peered out as an observer through a small, heavy plate glass window, presenting sea creatures from a narrow, entirely human perspective, as in an aquarium or zoo.[27]

Rachel Carson changed all that. Using the power of imagination and the lure of narrative, combined with scrupulous research and the latest scientific findings, Carson fundamentally altered how we think about the ocean, its ecology, and its creatures. *Under the Sea-Wind* teems with life that we enter into, rather than observe. It is a world with phosphorescent diatoms, gleaming globules of eggs by the millions, and omnipresent predators — from copepods and ctenophores to owls, orcas, and sharks. To bring marine life and the ocean, with all its diversity and ecosystems, fully alive and not merely enumerated as a resource for the amusement, edification, or enrichment of human beings, is, for Rachel Carson, a fundamental, conscious decision.

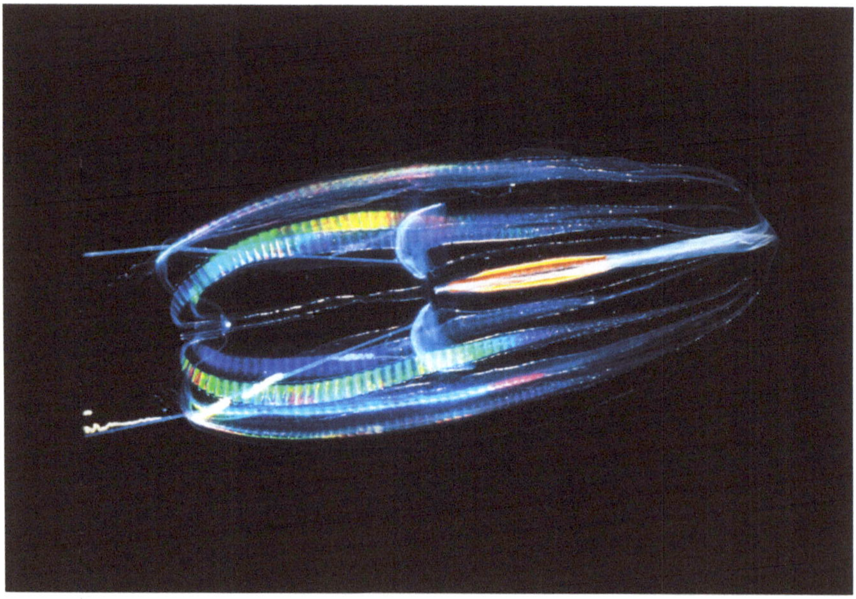

Rachel Carson's Hopes for Writing

Rachel Carson made her intentions for *Under the Sea-Wind* quite clear, both in her Foreword and in a lengthy memo to a member of the marketing department of Simon & Shuster, someone appropriately named Mrs. Eales.[28] In her Foreword, as if having Hendrik Van Loon in mind, Carson says, "*Under the Sea-Wind* was written to make the sea and its life as vivid a reality for those who may read the book as it has become for me during the past decade."[29]

To Mrs. Eales, Carson explains that she did not grow up near the sea; but it had long fascinated her, as she tried to imagine its appearance, the sound of its surf. Until she went to the Marine Biological Laboratory at Woods Hole, she had never seen the ocean, nor known first-hand its wondrous life forms. There she experienced "shore birds and fishes and beach crabs and all the other creatures that live in the sea or along its edge."[30]

We can feel Carson's early wonder and excitement, the vividness of her trips in a dredging boat up and down Vineyard Sound or Buzzards Bay. "After a time, with much violent rocking of the little boat, the dredge would be pulled up and its load of sea animals, rocks, shells, and seaweeds spilled out on the deck. Most of these animals I had never seen before; some I had never heard of. But there they were before me, dripping with sea water...Probably that was when I first began to let my imagination go down through the water and piece together bits of scientific fact until I could see the whole life of those creatures as they lived them in that strange sea world."[31]

Sea Creatures as Sentient Beings

Rachel Carson wants her readers to undergo a radical shift in perspective, away from the idea that humans are central to life on earth, to see and feel for themselves that other species are alive and sentient. They depend, as do we, on the bounty of our world for sustenance and survival. "I wanted my readers to feel that they were, for a time, actually living the lives of sea creatures. To bring this about I had first, of course, to think myself into the role of an animal that lives in the sea."[32] To achieve the effect of becoming another animal, Carson believes, requires not only active imagination, but "the temporary abandonment of human concepts and human yardsticks… time measured by a clock or the calendar means nothing if you are a shore bird or a fish, but the succession of light and darkness and the ebb and flow of the tides means the difference between the time to eat and the time to fast, between the time an enemy can find you easily and the time you are relatively safe."[33]

This shift in consciousness is achieved by Carson's assiduous attention to detail and magnified by her decision to anthropomorphize sea creatures, give them names, and describe their actions in terms analogous to human perceptions and feelings such as fear. We now know that Carson's observations, her use of the latest research from the Bureau of Fisheries and elsewhere, and her imagination allowed her to intuit and describe much about fish and bird behavior that has been shown by contemporary biologists to be grounded in fact.[34] *Under the Sea-Wind* is the progenitor of an important and popular contemporary genre of science writing. It includes Rachel Carson's descendants like Sy Montgomery, Ed Yong, Carl Safina, Jennifer Ackerman, and Diane Cramer.[35] They accurately and eloquently describe the amazing abilities of animals to think, feel, and communicate, making them worth caring about as individual beings, not simply objects of human sustenance or sport.

Seeking to present "the whole life of those sea creatures as they lived them," Rachel Carson is the first popular science writer who chooses to create feeling, smart, and sentient beings of fish and birds whose senses – sight, smell, taste, touch, hearing – allow them to anticipate danger, panic when trapped, and plan how to best attract and catch their prey. In the opening of *Under the Sea-Wind*, for example, Rynchops, the Black skimmer, cuts the water with his bill, creating sound waves that attract to the surface small fish like blennies who believe they are hearing the sound of shrimp, one of their favorite foods. Rynchops then simply scoops up his dinner. As Carson puts it, "in the fish world many things are told by sound waves."[36] In the 1930s, as Carson was researching and writing, whether or not fish could actually hear was still the subject of experiments in laboratories and debates in scientific literature.[37] Carson would have been aware of those studies that were concluding that — despite their lack of ears — fish could, indeed, hear. As a writer steeped in literature, she would also have given credence to a lengthy tradition of anecdotes and non-scientific descriptions of their auditory ability from at least the time of Izaak Walton's *The Compleat Angler* (1653).[38]

Similarly, whether fish suffer pain and are truly sentient was in dispute. But, as shad are trapped by gill-net fishermen in the beginning of *Under the Sea-Wind*, we are horrified to learn that they do feel "something that was like a burning, choking collar."[39] Carson,

who had previously studied and written about shad, describes them migrating by some form of "memory" that is comprised of "heightened senses as the delicate gills and the sensitive lateral lines perceived the lessening saltiness of the water and the changing rhythms and vibrations of the inshore waters."[40] The trapped shad are then attacked by eels who are also guided by taste and smell. They "caught the taste of fish blood...slipped out of their holes and followed the taste trail through the water to the net."[41] To complete her picture of remarkably sentient fishes, Carson follows the entire life cycle of a single fish, Scomber, the mackerel, from embryo to mature male. As he develops, organs emerge and senses and sensibility with them. For the first time, Scomber feels "the touch of the sea." His developing eyes become "deep blue with pigment, and now it may be that they sent to the tiny brain the first messages of things seen."[42] As he migrates, Scomber grows and finds himself among potential predators; he knows for the first time, "the fear of the hunted." Amidst a school of voracious anchovies, who are in turn attacked by bluefish, Scomber first tastes blood, "a strange new taste came with the rusty color and was drawn in by Scomber with the water he passed through the mouth and gills."[43] But our hero fish does not merely perceive blood and predatory fish, he is affected emotionally. "Scomber, lying a foot under water with fins aquiver, felt with suddenly taut nerves the thin staccato vibrations of the racing launce and the heavier roll of the pursuing whiting."[44] Later, in the presence of a seine fishing boat, Carson says "the mackerel were nervous and uneasy."[45] They are clearly able to anticipate danger from inanimate objects. This ability to associate shapes, sounds, and light with more abstract concepts, like danger or dinner, helps Scomber survive. What otherwise would be seen simply as tiny lights in the water, is recalled and processed as shrimp to be tracked and eaten, "following the little darting lights of the shrimp, the mackerel easily found and captured as many of them as they could eat."[46]

Before writing *Under the Sea-Wind*, Rachel Carson had also studied and written about the mysterious eel.[47] Only recently have scientists solved some of the mysteries of eel migration, where they go, and what guides them on a long, perilous journey to the Sargasso Sea. The exact migration route remained a mystery to Carson and other scientists of the 1930s. Only in the last decade has the path of the eels been mapped and their ability to detect magnetic fields been measured. But Carson attributes to Anguilla, the eel, the protagonist of the last section of the book, senses like smell that contemporary marine scientists suspect are essential parts of eel migration and behavior.[48] Anguilla begins her journey in a pond in North Carolina and then navigates small streams before heading to the sea. As she moves, she is able to distinguish varied and entirely new landscapes: "with her keen senses she savored the strange tastes and smells in the water. They were the bitter tastes and smells of dead and rain-soaked autumn leaves, the taste of forest moss and lichens and root-held humus."[49]

As Anguilla reaches the sea, we encounter numerous other fishes and creatures who are both sentient and scary. Carson describes the submerged wreck of a fishing boat, *the Mary B.*, that hides a large octopus which "concealed itself" and then waits patiently for trout to swim by. It shoots out a tentacle to haul them in to its "waiting mouth" to be "torn apart "by the cruel beak, shaped like a parrot's."[50] But perhaps even more scary, conscious, and crafty than the octopus — now well-known for its intelligence — is the angler fish that Carson names Lophius. On the sea floor, Lophius disguises his bellows shape and wide mouth filled with sharp teeth. His wrinkled skin makes him

resemble a rock. Only two evil eyes and "a curious wand" like a fishing rod with a flap of flesh at its tip for a lure are visible. Curious trout approach the lure and are soon devoured. But Lophius is not sated and moves on; he soon is able to detect the presence above him of eider ducks at the surface for "the scent and taste of duck were strong in the water that passed over the taste buds covering his tongue and the sensitive skin within his mouth."[51] Lophius swims quietly up to the surface and with his huge mouth and teeth pulls under and kills a struggling eider. Though horrifying, such predation, by octopuses, angler fish, and others is presented as highly evolved in a world where sea creatures plot, plan, think, feel, and can be as scared as is the reader. Marine species in this natural, watery world have evolved together and are marvelously adapted to their environment.

In *Under the Sea-World*, only humans upset this ecological balance. It is humans who are the intruders with their technology that allows them to destroy large numbers of other species. After Lophius fills his huge belly with trout and even an eider, he retreats to the *Mary B.* where he "would lie in torpor for several days, digesting his meal."[52] Not so the human predators who show up when huge numbers of fish — scup and sea trout, cod and croakers, flukes and flounder, sea bass and many more — head further out to sea over the continental shelf to escape the arrival of winter and reach warmer water near the Gulf Stream. The humans now await them in specially-designed boats, large trawlers, that are "squat and ungraceful of line."[53]

Our identification with Scomber and Anguilla and the other creatures previously thought merely instinctive and insensate now radically shifts our perspective. It is feeling individuals who are being scoured up in the thousands by the mechanical, impersonal nets of the trawlers. Carson lets us know that "only a decade before," these fish had been safe "from the fishermen's nets once they had left the bays and sounds." Then, with new technology, the boats began to range more widely, dragging nets deep in the water to search for fish heading further and further south. Finally, the distant wintering grounds of their quarry were discovered. "From that time on, the trawlers came every season and took millions of pounds of fishes each year."[54]

Carson Connects Ecosystems

In revealing the living world of the ocean, Rachel Carson wants her readers to understand the ocean, like the land, is made up of vastly differing zones and ecosystems. In order to give a comprehensive, yet connected view of this little-known world, Carson divides *Under the Sea-Wind* into three different parts or "Books" — each bustling with life and then linked as a single narrative— the shore, or edge of the sea, with its birds, animals, and crustaceans; the ocean with its pelagic birds, mackerel, mullets, and sharks of many kinds; and the deep with its mysterious creatures like octopuses, giant squid, and strange species that lurk on the bottom, beneath rocks, and in caves.

Book One, "The Edge of the Sea," begins with the Black Skimmer, Rynchops, who enters mysteriously out of the evening shadows. "Before the moon had come to the full Rynchops had remembered the island." He is joined by other skimmers, tired and hungry after a long journey from the Yucatan, ready to rest, feed and then "lay their eggs and hatch their buff-colored chicks on the sandy islands of the sound..."[55] Rynchops roams the island throughout the evening, feeding and making strange harsh and barking cries. As we soar and search with him, we can see the shore below, then zoom in for close-ups of various creatures feeding and hunting along the surf, a shoreline alive with movement, sounds, smells, and predation. A rat sniffs the freshly-laid eggs of a Diamond-backed terrapin in the sand. "Snuffling and squeaking with excitement, the rat began to dig and, in a few minutes, had uncovered an egg, had pierced the shell, and had sucked out the yoke."[56] He would eat more, but hears a newly-hatched terrapin struggling out of the water onto the sand. The rat snatches the baby terrapin and carries it off. Engrossed while gnawing away its shell, he ignores the slowly rising tide that allows the approach of another predator. The rat, too, is overheard in the night, speared and eaten by a patiently waiting hunter, the Great Blue Heron.[57]

What might appear to a casual human observer as an empty evening beach comes alive with creatures who are attuned to even the smallest sounds. "It would have taken the sharpest of ears to catch the sound of a hermit crab dragging his shell house along the beach just above the water line: the elfin shuffle of his feet on the sand, the sharp grit as he dragged his own shell across another; or to have discerned the spattering tinkle of the tiny droplets that fell

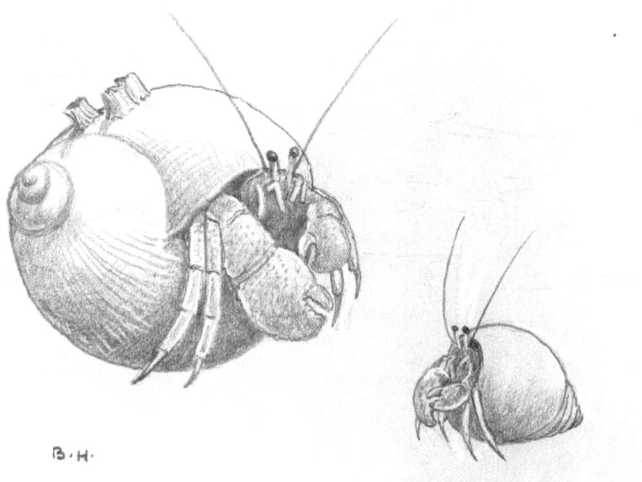

when a shrimp, being pursued by a school of fish, leaped clear of the water. But these were the unheard voices of the island night, of the water and the water's edge."[58] Giving "voice" to tiny, unnoticed creatures, piercing the silence, to human ears, of the world around us, are early, apt metaphors to which Carson will return often. If we are to understand our place in the world, we need to observe keenly and feel our connection to even the smallest life forms. "In every outthrust headland, in every curving beach, in every grain of sand there is the story of the earth."[59]

From Facts to Film: Carson's Sources Come Alive

Such attentive observation and detail, the gathering of scientific facts, and Carson's imaginative and dramatic ability to bring them to life give Book One the feel of a filmed, dramatic, marine documentary that had yet to be produced when *Under the Sea-Wind* was written, but which we now take for granted. Carson's cinematic narrative moves from moonlit night to the next morning in North Carolina where we meet a pair of sanderlings, Silverbar and Blackfoot, who hunt crabs, probing and running along the edge of the surf. It is May and spring migration. The sanderling are feeding, resting, gathering strength, as are curlews, plovers, and sandpipers of all kinds, before they set northward to the Artic, thousands of miles distant, where they breed and are beset by a succession of dangers in the struggle for survival.

Unlike contemporary science writers and documentarians, Rachel Carson was unable to visit the Arctic. She based much of the

detail that gives her writing such power on the life histories of birds assembled and published by Arthur Cleveland Bent in 1927. There are rich observations in Bent's volumes, but they are presented as bare scientific facts. With her pen, Carson fleshes them out, colorizes them and sets them in motion. We can see, hear and feel the mating displays and aggressiveness of the male sanderling, the plants used to line a sanderling nest, the female's scattering of eggshells away from her nest, drawing away the attention of predators like ravens and jaegers by making herself visible and vulnerable, or the surprised notation by one explorer that there are not more sanderlings lost to "Polar foxes" with their keen sense of smell.[60]

In "Arctic Rendezvous," the tundra where Silverbar builds her nest is transformed into a pastoral painting or photograph. "The rocks were clothed with mosses and soft gray lichens…There was a sparse growth of dwarf willow, with bursting leaf buds and ripe catkins. From scattered clumps of green the flowers of the wild betony lifted white faces to the sun, and over the south slope of

the hill was a pool fed by melting snow." The brief period of Arctic spring during which mating must occur comes alive with flowers, color, sounds: "first, the white cups of the mountain avens; then the purple of saxifrage, then the yellow of the buttercup glades, loud with the drone of bees trampling the shiny golden petals and jostling the laden anthers…the tundra was gay, too, with moving bits of color, for the midday sun coaxed out the butterflies from the willow thickets…" In this lush landscape, Red knots utter soft "coo-a-bees" and Lapland longspurs sing a "tinkling, many-syllabled song." But a polar fox invades this idyllic Arctic scene, padding silently, and, seeing Silverbar, "sniffed the air, believing she had young nearby." We sense Silverbar's devotion and desperation as she "watched the fox uncover the shells and nose them…" Then, in a selfless maneuver, she "fluttered toward him, tumbling to the ground as though hurt, flapping her wings, creeping over the gravel. All the while she uttered a high-pitched note like the cry of her own young. The fox rushed at her. Silverbar rose rapidly into the air, and flew over the crest of the ridge…"[61]

In Book Two, Carson again transmutes the dry, scientific facts from her scientific sources about the lives of fish into colorful, high

drama. Scomber, the mackerel, whose story takes up the middle of *Under the Sea-Wind*, begins life among thousands of eggs, many of which are devoured by voracious, tiny jellyfish-like animals called ctenophores. Then, as a young mackerel, he survives attacks by anchovies who are, in turn, slaughtered by larger bluefish. We realize that Scomber is in danger, is sentient, and is fearful; we can visualize, feel, even taste the rapid, chaotic action and attacks that surround him.

"In the surface waters Scomber first knew the fear of the hunted...Out of the clear green water a dozen gleaming silver fishes suddenly loomed up. They were anchovies, small and herringlike...Their silver scales flashed on all sides of him...A pack of young bluefish had picked up the scent of the anchovies...In a twinkling they were upon their prey, fierce and ravening as a pack of wolves...Two clean-severed heads and two tails floated away. The taste of blood was in the water...as the carnage spread, the clear green of the water was slowly clouded with a spreading stain. A strange new taste came with the rusty color and was drawn in by Scomber with the water he passed through mouth and gills. The taste was disquieting to a small fish that had never tasted blood or experienced the lust of the hunter."[62]

Moving up the coast during his larval and youthful stages, Scomber survives this and other assaults to emerge in the harbor at Cape

Cod in the "sea-coat" of a full-grown mackerel "sculptured to a torpedo-shaped body, sleek, swift, muscled, and colorful, but "soft as velvet to the touch." Out of the thousands of mackerel that reach the harbor each year (where Carson had observed them closely at Woods Hole) we are focused on, appreciate, and identify with this single, beautiful, sentient creature with whom we have swum since his fertilization, gestation and birth. "His back was a deep blue green — the color of the deep places of the sea that Scomber had not yet seen - and over the blue-green background irregular inky stripes ran from the back fin halfway down his flanks. His underparts gleamed of silver, and when the sun found him as he moved just under the surface of the sea he glittered with the colors of the rainbow."[63]

In Book Three, where we follow Anguilla, the eel, Carson again fleshes out and layers her oils over a canvas of flat, faceless facts about the mysterious migration and maturation of eels from the time they are embryos. She magnifies the eel embryos, making their surroundings shine, as if on the silver screen.

"Their small bodies were colorless as crystal. Even the blood that ran in its channels, pumped by hearts of infinitesimal size,

was unpigmented; only the eyes, small as black pinpricks, showed color...Before the eyes of the eels, clouds of copepods vibrated in their ceaseless dance of life; their crystal bodies catching the light like dust motes when the blue gleam came down from above. Clear bells pulsated in the water, fragile jellyfish adjusted to life where five hundred pounds of water pressed on every square inch of surface. Fleeing before the descending light, shoals of pteropods or winged snails, swept down from above before the eyes of the watching eels, their forms glistening with reflected light like a range of strangely shaped hailstones – daggers and spirals and cones of glassy clearness...Strangest of all were the fishes half as long as a man's finger and clothed in a leathery skin, that shone with turquoise and amethyst lights and gleamed like quicksilver over their flanks. Their bodies were thin from side to side and tapered to sharp edges. When enemies looked down from above they saw nothing, for the backs of the hatchetfish were a bluish black that was invisible in the black sea."[64]

Immersed as we are in the life of the ocean, Carson's poetic and painterly prose has upended our ethics entirely as we struggle to survive along with the creatures of the sea. When they appear, it is humans who are the intruders, humans whom we fear. When working at the Beaufort Fisheries Station, Rachel Carson watched a spectacular fish migration that had gone on undisturbed for eons — the running, leaping, and splashing of thousands of mullets to the sea. But now, this spectacular start of the annual mullet migration attracts fishermen in droves, waiting there to benefit from an easy kill.[65] Carson presents the scene pulsating with panic and colorful motion as the mullet first rush to the sea and then are trapped by their predators.

"The ebbing tide carried the mullet through the deeper green glooms and over the white sandy bottom of the channel.... Above them, as they moved, the surface of the water was broken into a thousand glittering facets that shone with the sun's gold... One after another the mullet rose to the shimmering ceiling of the sound.

One after another they flexed their bodies in a quickening rhythm, gathering their strength and leaping into the air."[66]

As they make their way through surf, driven by instinct to the sea, they are met by fisherman who have lain in wait, set their nets, and pursue the mullet in their boats.

"The shallow water in which the mullet are swimming is a pale, translucent green, murky with the loose sand which the waves are stirring up…A shadow looms in the green, sun-filled water in the path of the mullet. From a dim, gray curtain the shadow resolves itself into a web of slender, crisscross bars...As the first waves of panic pass from fish to fish, they dash shoreward, seeking a way of escape… They run seaward, but meet the circle of the net that is growing smaller, foot by foot, as the men on shore and in water up to their knees brace themselves in the sliding sand and pull on the ropes…"[67]

Under the Sea-Wind as Best-seller

Given such unprecedented, vivid, dramatic, and moving encounters with life in the sea, when *Under the Sea-Wind* was released in a new edition in 1952 by Oxford University Press, it quickly joined *The Sea Around Us* on the *New York Times* best-seller list.[68] This time, the reviews were even more ecstatic. Sterling North, in the *Washington Post*, surmises that Miss Carson holds the reader throughout fascinating lessons in marine science with "magic." He means that Carson is "not only one of the best informed individuals in the world on her subject," but that she "is one of the very few prose stylists now practicing who knows that there is only one possible word (of the right nuance of meaning, rhythm and verbal music) for its given place in a sentence."[69] The *New York Times*, which originally had been cool to *Under the Sea-Wind*, compared Rachel Carson to literary giants. "Great poets from Homer, with his sonorous hexameters on the 'loud-sounding sea,' down to Masefield... have tried to evoke the deep mystery and endless fascinations of the ocean; but the slender, gentle lady who is editor of the United States Fish and Wildlife Service seems to have the best of it. Once or twice in a lifetime does the world get a physical scientist with literary genius."[70]

But the most perceptive and important review of *Under the Sea-Wind* when it was republished came from the renowned, now classic, nature writer, Henry Beston, author of *The Outermost House: A Year of Life on the Great Beach of Cape Cod* (1928). It is also the review that gave Carson the most satisfaction; she had long idolized Beston and went out of her way while on Cape Cod to see his house and meet her hero.[71] Beston seems prescient about the effect Carson will have on our future, seeing her writing as a "portent," that an age "which has seen poetry die out of life has had an awakening and a reassertion of the human spirit and has

sought and welcomed a presentation of nature made valid by a sense of beauty and a superb power of poetic perception. Some spiritual instinct has shaken itself free, and has refused to take the scientific vision of nature as complete." Beston is the first to point out that in *Under the Sea-Wind* Carson calls into question what she would later call the "arrogance" of humanity's use of science to control nature and problematically put our species at the center of the universe. Calling the book "a very great achievement," Beston concludes that *Under the Sea-Wind* is "a book for all who do not fear the adventure of living. We shall live more fully for having read it." [72]

As additional marine science developed during World War II, it expanded knowledge about the ocean and Carson's understanding and love for it. Because of *Under the Sea-Wind*, Rachel Carson was able to make these startling new discoveries accessible to the public while giving them profound meaning; she connected human life and society to the vast and little-known ocean. She introduced much of the post-war American public to our shorelines and seas, to ecology and evolution, and, hence, to environmentalism. *Under the Sea-Wind*, *The Sea Around Us*, and *The Edge of the Sea*, Carson's trilogy of masterful best-selling works, astounded millions of admiring readers between 1951 and 1955, vaulting Carson into the pantheon of great American writers.[73]

Carson is now the ideal writer to produce *Silent Spring*, blending the vivid, poetic prose of her ocean books with increasingly sophisticated science in order to reach a huge swath of the public. Having seen her sensitivity and love for the lives of even mackerels and eels in *Under the Sea-Wind*, we understand why Carson is able to engage the emotions of her readers with a book on the health effects of persistent chemicals, one she called "her poison book." Those who recall reading *Silent Spring* often allude to its opening "Fable," a fictional account of an American town left empty and bereft of bird song. Carson's sense of justice is also on display as she relates the effects of pesticides on California farm workers.[74] Because of the mastery she had achieved in *Under the Sea-Wind*, Carson has the ability to make her readers care deeply about the fate of backyard robins dying agonizing deaths from DDT.[75] *Silent Spring*'s combination of evocation and exposé lifted Rachel Carson among the immortals; but it also eclipsed her divine ocean writing where her ability to entrance readers was born. Like life itself, Rachel Carson emerged from the sea.

Close attention to *Under the Sea-Wind* and its literary, scientific, and ethical power demonstrates that Rachel Carson and her work is best seen, studied and understood as an ecosystem – each part, each entity seen as interrelated with, essential to, the other. Rachel Carson is at once an observant, caring and compassionate individual; at the same time, she is a superb and lasting nature writer, and an exemplar of using sound science to alert the public to environmental dangers and consequences. We properly revere *Silent Spring*. But it and Rachel Carson cannot be fathomed in isolation. *Under the Sea-Wind* is a classic in its own right, a clear

and necessary antecedent to Carson's triumph with *Silent Spring.* In fact, only a few years after *Under the Sea-Wind* was first published, Carson learned about the potentially dangerous health effects of DDT from research by her friend and fellow birder, Chan Robbins, and colleagues at the Patuxent Research Refuge. Given her gifts demonstrated in *Under the Sea-Wind* and given encouragement, Rachel Carson could have begun *Silent Spring* as early as 1946. She attempted then to write about DDT for the *Reader's Digest.* But, to our loss, she was rejected.[77]

The Literary Rachel Carson

That Rachel Carson should be recognized as a great writer, a literary figure, and not merely for her exposé of the chemical industry in *Silent Spring*, was first suggested by Paul Brooks. Brooks was Carson's close friend and editor throughout her life. In his pioneering biography, *The House of Life: Rachel Carson at Work* (1972), he delineates Carson's views on writing and the writer, traces her literary influences, including important authors like Joseph Conrad who wrote about the sea,[78] and explains that Carson was never ashamed of her emotional responses to nature. "The experience I relate in *Under the Sea-Wind* of the young mullet pouring through that tide race to the sea is one that comes to mind...I didn't tell it as a personal experience, but it was – I stood knee-deep in that racing water and at times could scarcely see those darting, silver bits of life for my tears."[79]

Brooks was also the first to note Rachel Carson's affection for *Under the Sea-Wind* and its literary quality. Others, like Carson's friend, Judge Curtis Bok, a distinguished author and publisher of the *Saturday Evening Post,* felt similarly. "I think the *Sea-Wind* will always be my favorite – a quiet and special harbor where one stores up strength and peace before riding the wonders of Oceanus outside."[80]

Influences on Under the Sea-Wind

The "special harbor" that is *Under the Sea-Wind* is fed by several streams of Carson's literary antecedents in the Romantic era, in British nature writers, and the distinctively American nature study movement. Carson was deeply influenced by the popular British nature writer Henry Williamson whose books, *Tarka the Otter* (1928) and *Salar the Salmon* (1935), featured vivid description and anthropomorphism to draw readers into the lives of these creatures of the sea and shore. The opening of *Salar the Salmon*, like that of *Under the Sea-Wind*, is set in moonlight, with silver water shimmering with opalescence. It introduces us to the spawning and migration of small samlets from inland rivers to the sea, much like the transformation of tiny elvers that make their epic journey as eels to the Sargasso Sea.[81] The prose and plots of Williamson's books are so swift and dramatic that, even years later, Carson read *Tarka the Otter* aloud to her adopted son, Roger Christie.[82] Of Williamson, Carson wrote that his "nature writing was of the highest order."[83]

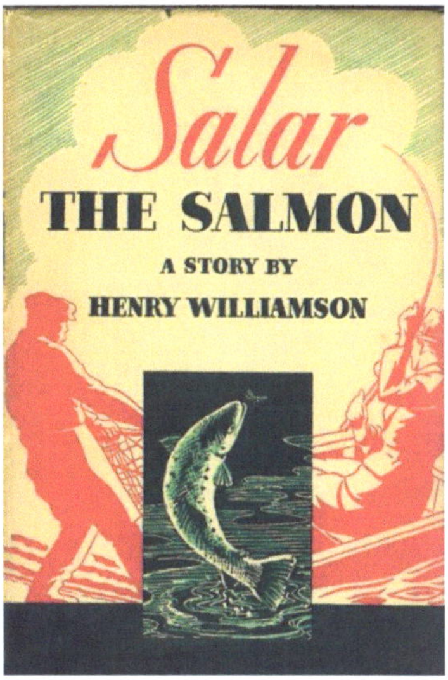

Rachel Carson often told how she was fascinated by the sea long before she ever saw it. Carson was drawn to it by literature's ability to transport us to places, times, societies, and characters far different than anything we have ever known. The great authors and poets stir our emotions and imagination so that we can see, almost touch, and feel deeply for strangers. It was at night, reading in her dorm room in the old Berry Hall with its leaded windows at the Pennsylvania College for Women, amidst a rag-

ing thunderstorm, that Rachel Carson became enraptured with the ocean through the words of the lovesick protagonist of Alfred, Lord Tennyson's poem, "Locksley Hall."[84] It ends with the young protagonist avowing, "Cramming all the blast before it, in its breast a thunderbolt. Let it fall on Locksley Hall, with rain or hail or snow; For the mighty wind arises, roaring seaward, and I go." Though "Locksley Hall" is perhaps best known for its line, "In Spring a young man's fancy turns to love," or its vision of a future world of peace and commerce, it is a seaside poem whose narrator reflects on his youthful musings from "yonder ivied casement."[85] Carson biographer, William Souder, worries that Carson was so taken, instead of appalled, by "Locksley Hall" which, some critics contend, is about a young man apparently about to join the British imperial military.[86] But Carson, the young female English major with a love of nature, would, instead, have been swept away by the romantic seaside setting of Locksley Hall. The poem opens in the early morning with the "curlews call" and "dreary gleams about the moorland" as it overlooks the "sandy tracts" and "hollow ocean-ridges roaring into cataracts." The protagonist wanders along the beach and sees a vision of the world as it could be and "all the wonder that would be." The springtime he imagines is precisely what Carson will later describe as lost to pesticides and progress in *Silent Spring*. "Locksley Hall" offers a hopeful vision where "a fuller crimson comes upon the robin's breast" and a "livelier iris changes on the burnish'd dove."[87]

Rachel Carson, who read her own drafts aloud until satisfied with the sound of their sentences, and who attempted unsuccessfully for years before *Under the Sea-Wind* to publish poetry,[88] is shaped by the sound and sensibility of nineteenth-century verse with its themes of the clash of death and life, survival and evolution, and the precious uncertainty of life in liminal spaces – especially the edges of the sea. Algernon Swinburne's poem, "A Forsaken Garden," [89] appeals deeply to Carson — its sounds, setting, and sentiment that life continues, despite the death of life and love. Swinburne describes a long-abandoned garden filled with dead weeds and brush that "fronts the sea" where a man "shall find no life but the sea-wind's, restless Night and day." Here the "nightingale calls not" with the only sound that of "the note of a sea-bird's song," much like Tennyson's curlew. Carson chose a line from this poem full of withered life as the epigraph for *Under the Sea-Wind*, "Till a last wind's breath upon all these blowing Roll the Sea,"[90] But its final stanza states more directly the central theme of *Under the Sea-Wind*. It is the death of Death, a god in terms of human life spans, but not in the face of geologic eons. What is eternal is the sea. Rising and falling throughout the ages, it is the sea, with its wind, that endures. As Carson puts it in her Introduction, "To stand at the edge of the sea, to sense the ebb and the flow of the tides, to feel the breath of a mist moving over a great salt marsh, to watch the flight of shore birds that have swept up and down the surf lines of the continents for untold thousands of thousands of years…is to have knowledge of things that are nearly as eternal as any earthly life can be. These things were before ever man stood on the shore of the ocean and looked out upon it with wonder; they continue year in, year out, through the centuries and the ages, while man's kingdoms rise and fall." [91]

The transitory nature of human civilization in the face of modern, scientific understandings of geology and evolution is at the heart of Carson's world view. By the time of *Silent Spring*, when humanity itself threatens life, Carson is drawn again to images and meaning from poetry. John Keats' line from "La Belle Dame Sans Merci," "The sedge is wither'd from the lake, /And no birds sing" was shown to Rachel Carson by her friend and literary agent, Marie Rodell. It finally settled the lingering question of the title for *Silent Spring* and became a powerfully resonant epigraph.[92] Ultimately, Rachel Carson sees Keats' poem as a metaphor for mankind's short-sighted and arrogant desire to control nature for our own ends. As literary critic John Elder puts it, "great poetry remained her criterion for vivid, organic wholeness." Carson's uses of poetry go deeper than skillful quotation and resonant references. "They also ground her," Elder says, "in a Romantic tradition of revolt against mechanical and hierarchical approaches to nature, and they convey the breadth of vision from which her prophetic power springs."[93]

As literature, scientific literature, *Under the Sea-Wind* employs a simple, but profound shift in point of view. Human beings are no longer the center; those whom we typically see as the "other" take the stage. We see and feel things as does a snowy owl mother trying to warm her eggs in a snowbound nest; we follow the miraculous growth of mackerel embryos from miniscule golden globules adrift on the surface of the sea to the sleek, streamlined, muscular fish whom humans catch in droves. It is this imaginative shift that undergirds all of Carson's great contributions, including *Silent Spring*.

Rachel Carson's Values, Maria Carson, Women, and the Nature Study Movement

Rachel Carson's ability to enter into the experience of other creatures and people comes not only from her immersion in poetry and literature. It also stems from the influence of her mother, Maria McLean Carson, who introduced Rachel to nature, science, concern for others, and the importance of ethical action. Maria McLean Carson, was the daughter of a learned Presbyterian minister, a brilliant student at the top of an elite women's preparatory school, and a teacher who drew on the experiential pedagogy and civic values of the Nature Study movement of the late nineteenth and early twentieth century. When Maria McLean married Robert Carlson, she was required, as was typical of the time, to give up her teaching job. Robert Carlson proved to be an inadequate provider and was rarely home at the small farm he had purchased east of Pittsburgh as he tried and failed at various ventures. And so, using the curriculum of Anna Botsford Comstock's *Handbook of Nature Study* as her guide, Maria poured all of her talent, education, and pent up dreams into her youngest, brightest, and favorite child, Rachel.[94]

The nature study movement and popular woman writers of the time, like Florence Merriam Bailey and Olive Thorne Miller, believed that exposure to the outdoors with a caring adult and learning to observe and love the birds, plants, and animals was superior to the dry, technical approach of the male scientists who shot and killed birds and other species so they could dissect and study them in the lab.[95] Always ambitious for Rachel, Maria helped her create her first book, ten hand-written pages with childish drawings, of her "animal friends."[96] Rachel roamed the fields and woods around home with her mother and, later, alone. She continued to observe nature, read prodigiously, and write. She won literary prizes from the popular magazine *St. Nicholas* that published the writings of Susan Fenimore Cooper, Emily Dickinson, and Edna St. Vincent Millay, and went off to the Pennsylvania College for Women (now Chatham University) to become an English major and a writer.[97]

MY FAVORITE RECREATION
BY RACHEL LOUISE CARSON (AGE 14)
(Honor Member)

THE call of the trail on that dewy May morning was too strong to withstand. The sun was barely an hour high when Pal and I set off for a day of our favorite sport with a lunch-box, a canteen, a note-book, and a camera. Your experienced woodsman will say that we were going birds'-nesting—in the most approved fashion.

Under the Sea-Wind and Science

But Rachel's success and strength are also founded upon her scientific accuracy, her wide-ranging knowledge, her use of scientific sources and scientific training. It was her mother who first introduced her to science through experiential learning as a child. In addition to her lifelong love of birds and animals, Carson's interest in geology, genetics, and the sea may have been stimulated by her discovery, on one of her childhood rambles, of a fossil fish.[98] Her sense that evidence of the emergence of life from the sea is all around us and is central to appreciating the species which have evolved since then can be seen in her field notes while observing raptors from landlocked Hawk Mountain in Pennsylvania. "Always in these Appalachian highlands there are reminders of those ancient seas that more than once lay over all this land. Halfway up the steep path to the lookout is a cliff formed of sandstone; long ago it was laid down under shallow marine waters where strange and unfamiliar fishes swam; then the seas receded, the mountains were uplifted, and now wind and rain are crumbling the cliff away to the sandy particles that first composed it. And these whitened limestone rocks on which I am now sitting – these, too, were formed under that Paleozoic ocean, of the myriad tiny skeletons of creatures that drifted in its water."[99]

Carson's sustained scientific training began under the influence of her charismatic PCW biology professor Mary Scott Skinker; it is the reason Carson switched her major from English to biology and gained a fellowship to Johns Hopkins graduate school and admission to the Woods Hole Biological Laboratory. Working her way through graduate school, Carson studied genetics with H. S. Jennings and Raymond Pearl, for whom she worked as a half-time assistant, taught biology as an adjunct at both Johns Hopkins and the University of Maryland Dental and Pharmacy School, and completed her master's thesis on "The Development of the Pronephros During the Embryonic and Early Larval Life of the Catfish (*Ictalurus punctatus*)."[100]

All of Carson's work, from *Under the Sea-Wind* to *Silent Spring*, is backed up by her training as a marine biologist, her exhaustive research in primary and secondary sources, her correspondence with leading scientists in the United States and globally, and by her friends and colleagues in the U.S. Fish and Wildlife Service where she was an aquatic scientist and senior editor. When *Silent Spring* was published, Carson was ridiculed and attacked by chemical corporations and their conservative allies who claimed that she was hysterical, a communist, and a woman without sufficient scientific standing. The attacks on Carson's scientific training and accuracy were clearly part of an orchestrated campaign to smear her.[101] But, even a cursory glance at *Under the Sea-Wind* would have dispelled the slurs she suffered.

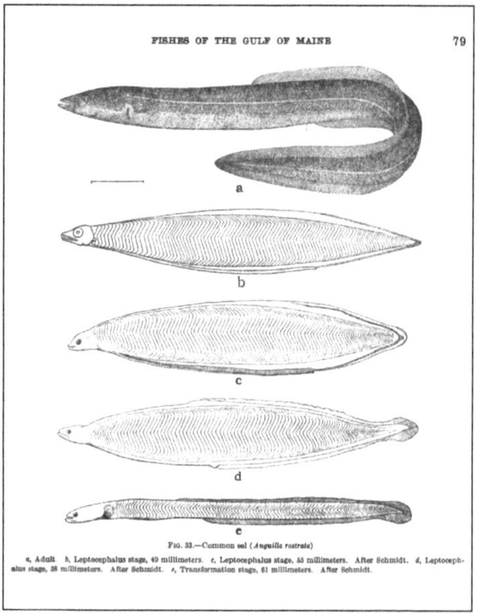

In the Foreword, Carson modestly notes that "no one person, even in a long lifetime, could become intimately acquainted through personal experience with every phase of the sea and its life...I have drawn freely upon a rich scientific and semipopular literature for basic facts upon which I have placed my own interpretation in weaving them into the story." Carson says that to "list all of the sources I have used would be impossible." But she goes on to list just a few "outstanding" examples, including Arthur Cleveland Bent's thirteen-volume life histories of North American birds, Henry B. Bigelow's *Fishes of the Gulf of Maine* (1924), Sir John Murray and Johan Hjort's "bible of oceanography," *The Depths of the Ocean*, (1912), and Johannes Schmidt's monumental papers on the life history of the eel.[102] These volumes alone contain encyclopedic histories of bird life and behavior, the development of marine exploration and oceanography, and are filled with detailed descriptions and black-and-white illustrations of species. Carson also mentions reading numerous journals and working with colleagues at the Bureau of Fisheries like her boss, Elmer Higgins, the ichthyologist, or Robert A. Nesbit, with whom she had documented marine life in the Chesapeake Bay, drawing on that work to provide numerous articles to the *Baltimore Sun* and other publications.[103]

70 DEPTHS OF THE OCEAN CHAP.

wished also to take a thoroughly good hydrographic section right across the Spanish Bay, with water-samples and tempera-

FIG. 48.—THREE SHORE FISHES FROM STATION 20, 141 metres (about 75 fathoms).
a. *Dentex maroccanus*, Cuv. et Val. Nat. size, 25 cm.
b. *Mullus surmuletus*, L. Nat. size, 29 cm.
c. *Peristedion cataphractum*, Cuv. et Val. Nat. size, 30 cm.

Silent Spring Eclipses *Under the Sea-Wind*

As a result of the immense impact of *Silent Spring* on the public, public policy, and the rise of the modern environmental movement, however, most of the attention to Rachel Carson over the years has focused on *Silent Spring*. Millions first saw Carson on television with an in-depth program with Eric Severeid on CBS Reports, *The Silent Spring of Rachel Carson*. It focused entirely on DDT and *Silent Spring*. In 1970, the first TV documentary on the Public Broadcasting System (PBS) took the same approach, included old CBS footage, and even used the title, *The Silent Spring of Rachel Carson.*[104]. During the 1980s, Carson became an icon for the early women's breast cancer movement as women marched waving copies of *Silent Spring.*[105] With the election of the pro-environment Clinton-Gore administration in 1992 came even greater revived interest in Rachel Carson's importance to environmentalism. But the emphasis was still on *Silent Spring* to which Vice President Al Gore wrote a glowing introduction for the 1994 edition.[106] Once again, *Under the Sea-Wind*, along with Rachel Carson's other ocean books, slipped beneath the waves.

A New Appreciation of Carson's Full Life and Work

As the environmental movement grew in size and influence during the Clinton-Gore years, it took Linda Lear's definitive biography, *Rachel Carson: Witness for Nature* (1997), to broaden the appreciation of Carson as a determined and often embattled woman, a writer whose life and work are marked by wonder, advocacy, and a love of the sea, represented by her influential and memorable trilogy of ocean books. Lear's reexamination of Carson shifted the view of her beyond the somewhat saintly and lone writer to create a portrait of Carson in the context of her entire work and life. By documenting the obstacles that Carson overcame including poverty, the lack of job opportunities for women in science, her family responsibilities and illness, Lear put Carson in a more complete, human light.[107]

Lear followed this fresh approach to Carson with an original collection of Carson's unpublished or hard-to-find writings called *Lost Woods: The Discovered Writing of Rachel Carson* (1998). *Lost Woods* published "Undersea," Carson's beautiful essay for the *Atlantic* that led to her contract for *Under the Sea-Wind*, the article on eel migration from the *Baltimore Sun* that was the precursor to the story of Anguilla in *Under the Sea-Wind*, and Carson's revealing long memo to Mrs. Eales at Simon & Shuster about how and why she came to write it.[108]

Carson's Centennial and Critical Interest

Lear's fuller interpretation of Rachel Carson, along with the growing influence of the environmental movement that saw her as its founding mother, set the stage for the 2007 centennial anniversary of Carson's birth. The centennial unleashed a tsunami of heightened interest in Rachel Carson. Linda Lear edited and wrote the Introduction to the 2007 centennial edition of *Under the Sea-Wind* and wrote the concluding, elegiac piece for a powerful collection of essays on Rachel Carson, *Courage for the Earth: Writers, Scientists, and Activists Celebrate the Life and Writing of Rachel Carson*, edited by the environmental writer, Peter Matthiessen.[109] This small volume shows deep appreciation, almost reverence, for Rachel Carson as a writer and an advocate. It includes contributions from Al Gore and other luminaries, as well as from a newer generation of writers directly inspired by Carson — Terry Tempest Williams on her moral courage, Sandra Steingraber on how activism created the possibility of *Silent Spring*, and Janisse Ray, beginning with "A Fable" of her own that features a town smothered in plastics. Ray shows how Carson's work has led us to understand the health effects of even tiny exposures to endocrine disrupting chemicals.[110]

The centennial also stimulated new public and scholarly discussion and evaluation as critics and academics sought to weigh Carson's place in history. Mark Lytle's short biography, *The Gentle Subversive: Rachel Carson, Silent Spring, and the Rise of the Environmental Movement* for Oxford University Press, emphasized Carson's political engagement and her influence on the environmental movement that followed. Unlike previous analyses, Lytle discussed and attempted to lay to rest the growing attacks on Carson and her work, including groundless claims that she was responsible for massive deaths in Africa because of restrictions on DDT. These attacks had begun in the nineties while Clinton and Gore sought to address global climate change and expanded environmental protection, especially in a new treaty internationally banning the chemicals Carson had exposed in *Silent Spring.*[111]

Two scholars, Lisa Sideris and Katherine Dean Moore, produced an influential collection of essays assessing Carson's overall work and standing in *Rachel Carson: Legacy and Challenge* (2008). They assembled a wide-raging group of authors to reflect Carson's ability to cross and combine disciplines and her distaste for and sense of the dangers of over-specialization "in an era of specialists." They explicitly argued for more attention to the whole of Carson's work, especially her ocean books, stating that, "It is difficult to overestimate the importance of *Silent Spring* in shaping our environmental consciousness. Yet, profound as that impact was and is, *Silent Spring* is only one part of Rachel Carson's legacy. Carson produced three major and highly acclaimed books"[112]

Later, in an influential essay in the *New Yorker*, historian Jill Lepore asserted that a new edition of *Silent Spring*, edited by ecologist and activist Sandra Steingraber, with additional pieces by and about Carson, wrongly neglected to include any of Carson's ocean writing. "It's heartbreaking to see that a new collection, *Silent Spring and Other Writings on the Environment*, edited by Sandra Steingraber (Library of America), includes not one drop of her writing about the sea." Lepore gave the mistaken impression that Steingraber did not appreciate or like Carson's ocean writing and was unaware that she was already working on a second complete edition of Rachel Carson's ocean books for the Library of America. But her keen observation about the importance of Carson's marine research and writing helped further revive interest in the ocean trilogy. "She [Carson] could not have written *Silent Spring* if she hadn't, for decades, scrambled down rocks, rolled up her pant legs, and waded into tide pools, thinking about how one thing can change another…"[113] Linda Lear also weighed in, urging greater attention to Carson's ocean writing, saying, "Ms. Lepore's point is well-taken," explaining that Carson's ocean trilogy is by "an ecologist, concerned with the relationships of organisms to their environment, long before that field was professionally recognized."[114] Then a new edition of *The Sea Around Us*, with an introduction by Sylvia Earle, came out urging attention to Carson's marine works. Earle, who explored the ocean as Carson was unable to, describes changes in our knowledge of the sea and science that Carson could not have known in the late 1930s. But Earle adds, "most remarkable to me is what she did imagine. Her writings are so sensitive to the feelings of fish, birds and other animals that she could put herself in their place, buoyed by the air or by water, gliding over and under the ocean's surface. She conveyed the sense that *she* was the living ocean…" Finally, the literary icon Margaret Drabble put out new single editions of the ocean books including *Under the Sea-Wind*.[115]

Under the Sea-Wind in Today's Environment

Read today, *Under the Sea-Wind* remains a superb introduction to the wonders and mysteries of the sea for non-scientists, using the best marine science available at the time. Although Carson continuously expanded and updated her scientific knowledge to produce *The Sea Around Us*, *The Edge of the Sea*, and the new Preface to the 1961 edition of *The Sea Around Us*, she did not have time to update the science in the 1952 edition of *Under the Sea-Wind* that had been rushed back into print to capitalize on the tremendous sales of *The Sea Around Us*.[116] Were Rachel Carson alive today, she could have updated what in the 1930s was cutting-edge oceanography. We have, for example, an entirely different understanding of the ocean's role in providing oxygen to the planet. Tiny phytoplankton the width of a strand of hair, were thought to be at the bottom of the food chain and the main source of oxygen through photosynthesis. Now with electron microscopes, lasers, and other instruments unavailable in Carson's day, we know that far

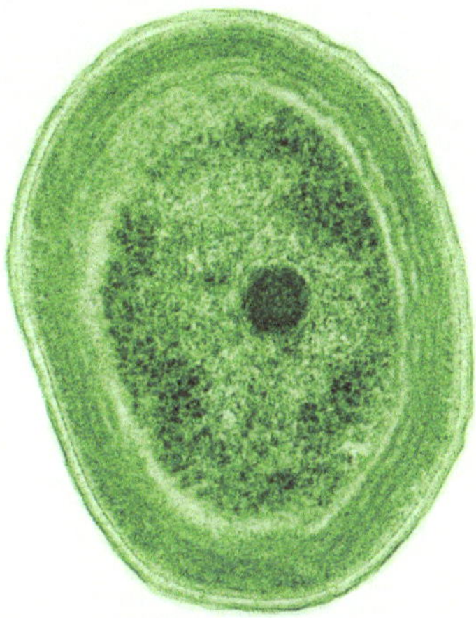

smaller, single-cell *Prochlorococcus*, of which there are about 150 million in a bucket of seawater, is the most numerous plant in the sea and produces about 20 percent of the oxygen in our atmosphere.[117] Carson would have found a way, as well, to note the dramatic changes for the worse in the ocean of today — from climate change, overfishing, pollution, species decline and loss, and more. She would have been devastated that Blackfoot and Silverbar, the sanderlings who migrated to the Arctic and survived bitter cold and snow, could now encounter June temperatures in the Arctic exceeding 100° F.[118]

Climate Change

Rachel Carson wrote about climate change in *The Sea Around Us* and was familiar with a Conservation Foundation report in 1963 warning about CO_2 levels and rising temperature by her friend and colleague, Roger Revelle, Charles Keeling, and other scientists.[119] At that time, CO_2 concentrations in the atmosphere were only 315 parts per million (ppm) when first measured accurately by Keeling at Mauna Loa in Hawaii where records that show the rising "Keeling Curve" are still maintained. Today, CO_2 has passed the dangerous level of 415 ppm.[120] Because of global climate change, average temperature in the Arctic has doubled, an increase of over 4° C. since *Under the Sea-Wind* was a best-seller in 1952.[121] Global ocean acidity from absorbed CO_2 is now 30% higher, threatening corals, mussels, crabs, and other shelled sea animals Carson described.[122] Sea-level rise has been growing about 1/8 of an inch per year, but on the coasts from North Carolina to Florida, the sites of Carson's observations, research, and writing in *Under the Sea-Wind* and the *Edge of the Sea*, the ocean has risen a full 5 inches between 2011 and 2015 alone.[123] The warming of the oceans means that between 10-60 per cent of embryos from spawning fish that depend on cool waters will be destroyed.[124] The tiny globes being eaten by ctenophores from which Scomber escaped to become the hero of *Under the Sea-Wind* will be gone. As a result of warming, overfishing and pollution, there are three-quarters fewer mackerel, tuna, and similar species since the 1970s.[125]

Pollution and Plastic

As we live and identify with the creatures under water in *Under the Sea-Wind*, we encounter little evidence of today's ocean pollution, including the modern scourge of marine species, plastic. Today, 73% of all marine debris is from plastics, with 8 million tons per year entering the ocean from land sources alone. Three hundred-sixty million tons of plastic per year degrade finally into tiny microplastics that are ingested by marine birds and animals. With reductions in sea ice, microplastics have recently been reported even in the depths of the Arctic Ocean.[126] Having absorbed microplastics, hermit crabs, who scuttled on the beach in *Under the Sea-Wind*, can no longer properly choose suitable shells to live in.[127] In the 1930s, the powerful neurotoxicity of mercury was not well understood; now there is three times as much mercury in the surface of the ocean.[128] When Rachel Carson was about to begin *Under the Sea-Wind*, William Beebe astounded the public in 1934 with a world-record Bathysphere dive of 3,028 feet; methyl mercury pollution has now been reported at the bottom of the ocean abyss to a depth of 36,000 feet.[129]

Anguilla Endangered

By the twenty-first century, the combined effects of overfishing, pollution, parasites, and climate change are so profound that Anguilla, the Atlantic eel, whose epic migration and life story concludes *Under the Sea-Wind*, is endangered. Catches of eels in American fisheries at the end of the twentieth century had declined by 75 per cent. Even in the once far off and mysterious Sargasso Sea, science writer Deborah Cramer reports, "Migrating eels bring in Asian parasites along with the chemical discards of a consumer society…Traces of chlordane and dieldrin, insecticides applied on land, lie at the bottom of the Sargasso…oil seeps in and coagulates into soft, sticky clumps of tar…Hatchling loggerhead turtles drift in the Gulf Stream at the edge of the Sargasso…The jaws of almost half these baby turtles are caked with tar…15 percent of the baby turtles have also ingested pieces of plastic or synthetic fiber."[130]

Although many marine species had been sharply reduced in the late nineteenth and early twentieth century, as Rachel Carson had reported in an early article on shad fishing, far more devastation was to follow with the industrialization and globalization of fishing in the late twentieth century.[131] *Under the Sea-Wind* allows us to feel what the ocean and our coasts were like when, compared to today, they teemed with abundant life. Reading it is a bit like standing with John James Audubon as he marvels at the skies darkened with billions of passenger pigeons stretching for miles, unsuspecting that this prolific and iconic American species would be extinct within the span of a lifetime.[132] Knowing the immensity of our environmental losses since the 1930s, Rachel Carson would surely have taken action against such devastation, would have spoken out to protect shores, seas, and the species who live in them. At first, Carson was a reluctant warrior. *Under the Sea-Wind* has no overt political implications, in contrast with Carson's later withering critique of the Eisenhower Administration's destruction of conservation leadership within government agencies,[133] her blistering attack on her critics and corporate interests in her speech to the National Women's Press Club,[134] or her condemnation of efficiency and greed in her attack on factory farms in *Animal Machines*.[135]

Rachel Carson and Politics

Some biographers have either downplayed Carson's advocacy and political engagement, as did Paul Brooks by stressing her literary strengths, or denied it altogether like William Souder who believes Carson's focus on the prose of her nature writing hero Henry Williamson, while failing to mention his pro-Nazi stance in the late thirties, marks her as essentially non-political.[136]

Rachel Carson did not merely oppose Dwight Eisenhower's stacking of conservation agencies with businessmen, she strongly favored Ike's Democratic opponent Adlai Stevenson and went on to campaign for and join women for John F. Kennedy. Working closely with the Kennedy Administration, as historian Douglas Brinkley has carefully described, Carson wrote speeches, drafted legislation, and actively pushed measures with Interior Secretary Stewart Udall to establish a conservation department that ultimately led to the EPA.[137] Much of the historic environmental legislation that emerged

after her death in 1964 is also properly credited to her influence since it was written or introduced by her many colleagues, admirers, and those she influenced. The Endangered Species Act of 1970, for just one example, protects not just species, but entire ecosystems, a concept whose widespread adoption flows directly from Carson's work.[138]

But even Rachel Carson's important and extensive legislative and electoral political activity posits too narrow an understanding of the term political and the impact and influence of her ocean writing, especially *Under the Sea-Wind.* Her first book radically shifts our perspective, our ethical stance, and our feeling for other creatures and their intrinsic worth, no matter how small or bizarre. Such achievements surely undermine a political system that values profit, property, and commerce above all else. Our horror at the struggles of Scomber, the mackerel, trapped in fishing nets and near death in *Under the Sea-Wind*, forever changes our view of fishing, especially as it has evolved into a massive, global industrial enterprise. The lyrical prose, the poetry and painting, the drama of Rachel Carson's first book that allows us to become a sanderling evading an Arctic fox, or an eel that today is critically threatened by our arrogance and greed, are intrinsically political and radical. It is why, to this day, Carson is not only revered, but attacked and vilified. Rachel Carson understood the political implications of her writing and ecological stance while she was still alive. It is why, as she neared death from metastasizing breast cancer and other health complications, she asked her closest friends and colleagues to form an organization to carry on her work that became the Rachel Carson Council.[139]

Carson's Living Legacy: The Rachel Carson Council

Following Carson's wishes, and led by Marie Rodell, Shirley Briggs, Irston Barnes, and others, the Rachel Carson Council was founded in 1965 and took up Carson's work amidst the continuing impact of *Silent Spring*, focusing on the effects of pesticides on wildlife and people.[140] As her executor, Marie Rodell saw to Carson's literary reputation, rights and sales, including the popular posthumous publication of *The Sense of Wonder*, a small book laden with photos that showed mothers how to introduce their children to the wonders of the natural world.[141] Shirley Briggs became the first long-term head of the Rachel Carson Council which was housed at the DC Audubon Society where Carson had served on the Board of Directors and, with Briggs, wrote for its newsletter, *The Thrush*, which became the magazine, *The Atlantic Naturalist*.[142]

Under the leadership of Shirley Briggs, the RCC provided expert information on pesticides culminating in the publication in 1992 of the definitive work, *Basic Guide to Pesticides*, that comprehensively detailed the pesticides and herbicides in use and their health effects.[143] Briggs was succeeded by Dr. Diana Post, a veterinarian, who, with her husband, Clifford Hall, purchased Rachel Carson's home in the Quaint Acres section of Silver Spring, Maryland, which Carson had designed and was built for her after the commercial success of *The Sea Around Us*. It is where she wrote *Silent Spring* and *The Sea Around Us*. Diana Post housed the Rachel Carson Council there, opened it to the public, and focused the work of the RCC on educating the public, homeowners, and pet owners on the dangers of using pesticides and other chemicals in everyday life. The RCC also promoted the legacy and reputation of Rachel Carson with conferences, anniversaries, and materials about her life and work, including marking her birthday, May 27, as official Rachel Carson

day with various cities and towns nationwide. Eventually, however, as the rise of the internet made information about pesticides and their effects easily and widely available, and as Carson's definitive biographer, Linda Lear, lectured widely, spoke to the media, created Carson archives at Connecticut College, and a website that offered authoritative information and resources about Carson's life and achievements, the central role and financial support for the Rachel Carson Council dwindled. Diana Post then retired from the RCC and carried out more local public education from the Rachel Carson house that became a non-profit entity called the Rachel Carson Landmark Alliance.[144]

Then, in 2014, after organizational collapse, the Rachel Carson Council was reborn when it brought on its current President & CEO, Dr. Robert K. Musil. Musil, an author, educator, and long-time national leader in movements for peace, justice, and the environment took up the challenge to build an organization that would reflect Rachel Carson's multiple, interconnected roles — as an advocate for environmental issues far broader than pesticides, including global climate change, as a writer who believed that awe, imagination and wonder were essential to developing empathy for all living things, and as a scientist who used her literary gifts to communicate complicated, controversial scientific findings to the general public and policy makers so they would take action. Having taught at a number of colleges and universities, lectured on campuses nationwide, served on a college's Board of Trustees, and published two highly-regarded books with the Rutgers University Press, *Hope for a Heated Panet: How Americans Are Fighting Global Warming and Building a Better Future* and *Rachel Carson and Her Sisters: Extraordinary Wom-*

en Who Have Shaped America's Environment, Musil was also in a position to build a nationwide campus network for the RCC that would educate, engage, and support the next generation of environmental leaders and allow students, faculty, and administrators to have an outlet to act on their environmental concerns.[145]

And, unlike the previous iterations of the Rachel Carson Council, Musil believed that carrying on Carson's work would mean that the RCC should be political in the broadest sense of the term, actively involved in organizing and advocacy in Washington, in key states, and at the grassroots. Within the first two years of the new Rachel Carson Council, the RCC held a live-streamed national conference on Capitol Hill featuring Senator Sheldon Whitehouse (D-RI), the leading champion in the Congress on global climate change, and newly-elected Senator Chris Van Hollen (D-MD) along with leading environmental writers and advocates, the heads of national environmental and peace NGOs, and environmental justice and grassroots leaders central to movements in the South opposed to factory farms, toxic chemicals, and global climate change.[146] Rachel Carson had once again made her mark on Capitol Hill. Senators like Tammy Baldwin (D-WI) told the RCC that Carson was her hero, while Sen. Tom Udall (D-AZ), the son of Secretary of the Interior Stewart Udall, whom Carson had worked with in the Kennedy era, spoke at RCC events, worked with the Council on

new toxics legislation, and shared stories from his boyhood of his recollections of Rachel Carson.

With its commitment to activism and justice, as well as rising visibility and political influence, the RCC was able to attract a growing number of youth, especially on college campuses, so that the Rachel Carson Council Campus Network (RCCN), initiated in 2018, quickly grew to 35 colleges and universities and now engages some 100 campuses with more than 5,000 active students and faculty.[147] Then, in order to find, recruit, mentor, and financially support the next Rachel Carsons, the RCC began its RCC National Environmental Leadership Fellowship (NELF) Program with an initial cohort of seven fellows. At the same time, given its work in the South, especially in North Carolina, the RCC became eligible for and was granted its first Duke University Stanback Fellow. Today, the RCC is awarded fourteen Duke Stanback Fellows who work full time at the RCC each summer. And, each year, the RCC awards over 30 national fellowships which provide a $2,000 stipend, mentoring, and opportunities to publish, while RCC Fellows carry out environmental projects and campaigns, and attend the five-day RCC American Environmental Leadership Institute (AELI) in Washington, D.C.[148]

Given its renewed reputation and access to political leaders in Washington, along with a strong grassroots network of activists, academic leaders, and the commitment and energy of youth, the RCC was able to build campaigns around key issues like climate justice, opposition to factory farms, fossil fuel infrastructure, regenerative agriculture, food insecurity, and forest preservation. The Rachel Carson Council launched its first major campaign against the effects of huge hog farms that in North Carolina hold some 10,000,000 pigs whose emissions cause air and water pollution, exacerbate global climate change, and harm human health. The RCC then aligned with the NC Environmental Justice Coalition and other groups, joined in demonstrations and actions with the Rev. William Barbour's Moral Monday movement, and helped pack hearings in the North Carolina legislature to try to block legislation to prevent nuisance suits by residents like Elsie Herring whose home bor-

dered a CAFO. The RCC then published its first major report, *Pork and Pollution*. It detailed the horrors of pork production, showed how to become involved in efforts to prevent it, and featured Rachel Carson's powerful and prescient opposition to the earliest factory farms from her Introduction to *Animal Machines*, a pioneering exposé by Carson's friend and colleague, Ruth Harrison.[149]

***The modern world worships the gods of speed and quantity, and of the quick and easy profit, and out of this idolatry monstrous evils have arisen.*—Rachel Carson**

The RCC expanded its campaigning against CAFOs to include chicken factory farms in central North Carolina and particularly in the Eastern Shore of Maryland where over 30,000,000 chickens are housed in two predominantly Black counties where Frederick Douglass and Harriet Tubman had been enslaved. As in North Carolina, with its report *Fowl Matters*, the RCC emphasized environmental justice since the pollution and adverse health effects were centered on lower-income African-American communities and immigrant workers in the factories themselves.[150] This work not only directly carried on Rachel Carson's opposition to factory farms, it also reflected her early concern for environmental justice in *Silent Spring* where she laments the exposure of California farmworkers to herbicides and pesticides.[151] Or, similarly, in a San Francisco speech to physicians in 1963, Carson decried the fact that U.S. atmospheric nuclear testing and radionuclides from its fallout had been spread worldwide by wind and rain. The radioactivity ultimately entered the food chain in the Arctic and accumulated in the blubber eaten by innocent Inuit (Eskimo) and Lapp mothers thousands of miles away from the original explosions. Radionuclides then could be found in their breast milk and harm nursing babies.[152] Such compassion and concern for justice, such empathy, reflects, as we have seen, Carson's religious, Presbyterian upbring, her own appreciation of the uniqueness and value of each miraculously evolved species, and, importantly, from her close friendship and work with her friend, literary agent, and executor, Marie Rodell. For it was Rodell who collaborated closely with Carson on *Silent Spring*, while, at the same time, was the agent for and intimately involved with the conception, editing, and bringing to fruition of the first book, *Stride Toward Freedom* (1958) of a then little-known activist Baptist preacher, the Rev. Dr. Martin Luther King, Jr.[153]

These early Rachel Carson Council campaigns connected pollution and health issues to global climate change which Carson had known and written about as early as *The Sea Around Us* and learned about in depth from the work of her friend and colleague, Roger Revelle. It was Revelle who had arranged for an observatory on Mauna Loa in Hawaii to be built that gave the first definitive readings of the rising concentrations of CO_2 in the atmosphere. Then, with other scientists, Revelle sent a report to the Kennedy administration in 1963 warning of the dangers of global climate change.[154] Drawing on this legacy, the RCC then focused directly on the fossil fuel industry and its infrastructure, including opposition to coal, oil, and gas. The RCC's report *Blast Zone: The Natural Gas and the Atlantic Coast Pipeline – Causes, Consequences and Civic Action* detailed the dangers of pipelines such as the leaks, fires, and explosions and was based on the first-hand observations and investigations of RCC Fellows who joined in marches and protests along the proposed pipeline route.[155]

As the Rachel Carson Council continued to grow and expand its staff, the number of its RCC National Environment Leadership Fellows, and its coalition partners, it was ultimately able to take a leadership role in a truly nationwide campaign aimed at the clear-cutting of forests to produce industrial scale wood pellets which are then shipped overseas to be burned in former coal-fired utilities to produce electricity. Carried out mainly by two corporations, Drax and Enviva, this little-known, but rapidly growing source of air, land, and water pollution, forest devastation, greenhouse gas emissions, and adverse environmental health and justice effects stretches from North Carolina to the Gulf Coast and westward to California and the Pacific Northwest.

Drawing on its environmental justice work in North Carolina, the RCC first helped initiate the Southern Forest Conservation Coalition (SFCC) which included the Southern Environmental Law Center, the Dogwood Alliance, local branches of the Sierra Club, individual environmental leaders and experts, and environmental justice community leaders fighting clear cutting and wood pellet production factories where they lived. As the wood pellet industry continued to expand, so too did opposition to it. The Rachel Carson Council produced four major reports, sponsored riverboat cruises along the Cape Fear River to observe and learn about wood

pellet production, made wood pellet production the center of its annual Capitol Hill Advocacy Days, and engaged a new generation of environmental leaders each year at its American Environment Leadership Institute (AELI).[156] In 2025, the Rachel Carson Council brought together campaigners, experts, journalists, environmental justice leaders, and leaders of the opposition to wood pellet production from the UK, the home of the Drax Corporation, at the RCC National Wood Pellet Forum II in Washington, DC.[157] Through its wood pellet campaign, the RCC also engaged a committed core of members of Congress including Sen. Sheldon Whitehouse and Sen. Chris Van Hollen, along with Reps. Jamie Raskin (D-MD), Don Beyer (D-VA), and Valerie Foushee (D-NC) to raise the issue and introduce legislation to stem the tide of wood pellets. The RCC then produced a short documentary, "Wood Pellet Wasteland" to be shown at film festivals, college campuses, and special events nationwide, along with advertising in national print and online magazines and journals like *The New Republic*, *Mother Jones*, the *Nation*, and *Grist*. [158]

With the Rachel Carson legend and legacy reinforced and resonating widely, the RCC was able to weigh in credibly on matters beyond environmental advocacy that Carson would have championed, including threats to democracy through the autocratic abuse of the Presidency by Donald Trump. Carson had, for example, directly and publicly opposed the Eisenhower administration's firing of the civil service head of the U.S. Fish and Wildlife Service, Alfred Day, and his replacement by an unqualified pro-corporate leader.

Thus, the Rachel Carson Council supported the impeachment of Donald Trump during his first and second terms and opposed his firing of civil servants, dismantling of federal agencies, especially those committed to the environment like the EPA, NOAA, NIH and the National Cancer Institute, and establishment of the illicitly authorized Department of Government Efficiency (DOGE) under Elon Musk with the primary purpose of weeding out employees deemed likely to be disloyal to the President.[159] As the midterm elec-

tion of 2026 loomed, the RCC launched a national voter education and engagement campaign designed to encourage voting and increase turnout especially among youth on college campuses across the country.[160] Like Rachel Carson herself, these heirs to her legacy would not only vote, but be encouraged to hold rallies, canvass, and even become engaged with campaigns as Carson had for Adlai Stevenson and John F. Kennedy.

Throughout 2026, as the Rachel Carson Council carried out nationwide education, organizing, advocacy, events and ads, film, video and social media based on the vision and values of Rachel Carson, the legend and legacy of Rachel Carson continued to grow. At Yale, where Marie Rodell had chosen to donate Rachel's manuscripts, letters and other materials to the then brand-new, state of the art Beinecke Rare Book and Manuscript Library, the Beinecke was mounting an extensive public exhibition "Silent Springs, Windswept Seas: Rachel Carson's Environmental Vision."[161] Carson is also featured at the USFWS's National Conservation Training Center (NCTC). The NCTC occupies a 500-acre campus along the Potomac River with woods, fields, hiking trails, and buildings including dorm-style housing and a major educational building. Each year, thousands of federal employees are trained at the NCTC which also holds regular environmental educational programs and lectures for the public. Attendees also visit the NCTC Museum and Archives which features Rachel Carson in displays along with artifacts, including her typewriter, and the books from her personal library.[162]

Rachel Carson's presence and continuing legacy is also tangibly felt at her alma mater, Chatham University. Rachel was inspired by Tennyson at Berry Hall on the original Shadyside campus in Pittsburgh. But Chatham has since created an entire large, sustainable Eden Hall campus that is the home of the Falk School of Sustainability and Environment outside of Pittsburgh. The complex states that its mission is directly inspired by its distinguished alumna, Rachel Carson, '29. In addition to courses and programs in sustainability and the environment, the Falk School offers a full-tuition Rachel Carson Scholarship, the Rachel Carson Healthy Planet Award for outstanding high school students, and chooses Rachel Carson Visiting Scholars to be in residence during the summer.[163]

And, following years of planning, preparation, and fund-raising, the first museum entirely dedicated to Rachel Carson and her work, the Springsong Museum, was set to open in Silver Spring, Maryland,[164] not far from where Rachel lived and wrote *Silent Spring*, bursting into national acclaim after her appearance on Eric Severeid's CBS special documentary, *The Silent Spring of Rachel Carson*, and her bold televised testimony to Sen. Abe Ribicoff's Senate Subcommittee on Pesticides. Rachel Carson also has a place in the National Portrait Gallery where a bronze bust of her by artist Una Hanbury has been variously displayed including in the hall dedicated to famous American women.[165] Carson is memorialized in bronze, as well, as she sits reading along the waterfront at Woods Hole in Cape Cod.[166]

But as she was dying, Carson did not ask her friends and colleagues to ensure that she and her books were not forgotten. She asked them to carry on her work. That work included active advocacy, lobbying, and campaigns in coalition with multiple organizations who were in

the 1950s already beginning to form a movement that would emerge with astonishing new power after *Silent Spring*. Through friends, Rachel Carson had met the head of the Wilderness Society, Howard Zahniser, in 1946 and was an active member by at least 1950. As a result, she was involved one of the first major environmental lobbying campaigns of the time, an extended struggle to prevent the building of a massive dam that would flood and destroy the Dinosaur National Monument.[167]

After Rachel's death in 1964, her work has been continued not only by the Rachel Carson Council, but by a host of active environmental organizations that Carson was connected with, or recognize Rachel's influence, and have incorporated her name directly. Rachel was, for instance, close to and collaborated with David Brower, then the head of the Sierra Club, who escorted Carson, in pain and confined to a wheelchair, through the redwoods of Muir Woods and other sites she had longed to see. After her death, Carson left significant funds to the Sierra Club which to this day receives bequests through its Rachel Carson Society.[168]

Rachel Carson also left funds to another national environmental organization with which she worked closely, the Defenders of Wildlife. Founded in 1947 by Mary Hazell Harris and others as the Defenders of Furbearers, the group sought to protect fur bearing animals that were being trapped and destroyed through predator control. Carson joined Defenders of Wildlife as a member in 1950. Rachel was brought on to its Board of Directors in 1964 shortly before her death by Harris, by then the Defenders longtime Executive Secretary (Executive Director) and a leader in the battle to establish the Endangered Species Act. Carson's bequest was used to set up the Rachel Carson Memorial Wildlife Education Fund which took aim at the ravages of predator control.[169]

The environmental movement had burgeoned following the impact of Rachel Carson and *Silent Spring*, the first Earth Day in 1970, and an environmentally friendly Carter Administration. But the advent of the telegenic, genial, and highly conservative and anti-environmental president, Ronald Reagan, a former movie star and Governor of California, meant the environmental movement would need to grow rapidly and develop new strategies suited to the new age of mass media and marketing. The Defenders of Wildlife was at the forefront of this shift as their new Executive Director, Rodger Schlickeisen, an entrepreneurial leader with an M.B.A. from Harvard, soon moved the defenders from a national headquarters occupying a small brick house to a glittering glass building, created a massive mailing list and membership, and made concern for wolves and other wildlife a major national issue. When Schlickeisen retired in 2011, this modernized, major organization became one of the first big green groups to be headed by a woman, Jamie Rapaport Clark, who was directly and

deeply inspired by Rachel Carson.[170] Clark had been the Executive Vice President at Defenders of Wildlife, leading Capitol Hill lobbying, operations, and conservation programs. But before that, she was also the first woman ever to head the U.S Fish and Wildlife Service and instrumental in the Endangered Species Act. Clark was so inspired by Rachel Carson that she even named her daughter Carson and wrote eloquently about Carson and the importance of women environmental leaders.

Rachel Carson also left royalties from *Silent Spring* to the Maine Nature Conservancy which she had founded in order to protect a salt pond on the shores of Muscongus Bay. It was from this quarter-acre salt pond that Carson gathered material for her book *The Edge of the Sea* (1955) — kelp, sea colander, dulse, dogwinkles, periwinkles, brown-green knotted wrack and rockweed, blue mussels, green crabs and red, blue and green Irish moss. When the salt pond was threatened by development, she took action. On a spring day in 1956, Carson and a handful of concerned citizens gathered at the Wiscasset Inn on the Maine coast to discuss the threats they saw around them and consider what could be done. Every idea had promise, but none seemed likely to succeed. Then Rachel Carson spoke. "What about The Nature Conservancy?" she asked. "It is the only group I know which is doing something practical about actually preserving areas."

Today, The Nature Conservancy's Rachel Carson Salt Pond Preserve is located at that very spot, but additional sites were purchased and then linked as a series of preserved coastal marshes and wetlands that are key ecosystems for migratory birds. In 1966, TNC turned these sites over to the state of Maine. Then, in 1970, this collection of salt marshes, estuaries, and rocky shores was given to the

U.S. Fish and Wildlife Service and now forms the Rachel Carson National Wildlife Refuge that stretches for over 50 miles and receives over a third of a million visitors each year.[171]

Other national environmental organizations have also helped spread Rachel Carson's legacy from when she was alive until today. The National Audubon Society was part of the original campaign to ensure that *Silent Spring* would be met with enthusiasm and become a best-seller. Audubon's president at the time, Carl Buchheister, was one of many influential individuals to receive a proof copy of *Silent Spring*. As a result, the *Audubon Magazine* editor, John Vosburgh, and Roland Clement, staff biologist, immediately asked if they could publish a two-part excerpt. *The New Yorker*, which would publish the first part of its serialization of *Silent Spring* on June 16, agreed, presumably understanding that this large enthusiastic audience would only help in building excitement and readership for *Silent Spring*.[172] When Rachel Carson and *Silent Spring* were then hit with a wave of attacks from conservative, business, and pro-pesticide outlets, *Audubon Magazine* leapt to Carson's defense, publishing the full text of her speech, a devastating response to her critics. It was a Carson classic still available on the Audubon website.[173] Then, in 1963, the Audubon Society followed up by giving Rachel Carson its highest award, the Audubon Medal. With such close ties to Rachel, whom Audubon sees as the foremother of the modern movement, the organization, which now has over 1,500,000 members, initiated its Rachel Carson Award in 2004. The award is given annually at a glittering luncheon at the Plaza Hotel in New York City and honors American women whose work has greatly advanced conservation locally and globally. Honorees have included grassroots organizers, environmental justice leaders like Dorceta Taylor,

former EPA heads like Lisa Jackson and Gina McCarthy, and women leaders of environmental NGOs like Abigail Dillen of EarthJustice and Jamie Rapaport Clark of Defenders of Wildlife.[174]

Among her many talents and roles, Rachel Carson was a journalist first published as a child and medal winner in *St. Nicholas Magazine.* Later, she freelanced for the *Baltimore Sun* and had her career jump started by her brilliant article "Undersea" in the *Atlantic* which led directly to *Under the Sea-Wind.* Carson also wrote for popular magazines like *The Woman's Home Companion* and *Holiday*, science magazines like *Nature*, and the *New Yorker* where *Silent Spring* was first serialized.[175] In addition to being a best-selling and award-winning author of historic environmental books, Rachel Carson is also a pioneer environmental journalist who has inspired generations of those who came after her.

It is no surprise, then, that the Society of Environmental Journalists (SEJ), founded in 1990, has since 2008 given out a coveted Rachel Carson Book Award for the best environmental book of the year, along with numerous runners up. SEJ's more than 1500 nationwide environmental journalists select the winners which are announced at the annual SEJ national conference. Recent winners include Alden Wicker, *To Dye For: How Toxic Fashion is Making us Sick*, Dan Flores for *Wild New World: The Epic Story of Animals and People in America*, and the late Tatiana Schlossberg, granddaughter of John F. Kennedy and climate and science writer for *The New York Times*, for *Inconspicuous Consumption: The Environmental Impact You Didn't Know You Have.* [176]

Part of the spectacular success of *Silent Spring* came from a carefully planned campaign to get the backing of prominent Americans, especially women like Eleanor Roosevelt, Frances Perkins, Agnes Meyer, owner and publisher of *The Washington Post*, as well as the heads of national women's organizations who were invited to a power lunch in Washington in order to get the word about *Silent Spring* out to their huge constituencies and followers.[177]

Today, a similar set of influential women philanthropists has formed Rachel's Network and expanded Rachel's legacy and influence to include advocacy in the Capital, education and organizing, retreats for women's empowerment, and particular emphasis on connecting, recognizing, and supporting women of color and their work for environmental justice. Since less than 1% of foundation giving goes to women and girls of color and organizations led by people of color receive less than 1% of multi-year operational grants, the group created its Rachel's Network Catalyst Award in 2019. It gives awardees each year for three years $10,000 as a personal grant, $5,000 for a well-being grant, and a $10,000 to their organization. In 2025, the Catalyst Award winners included: environmental justice leaders like Latricia Adams of Memphis, Tennessee, founder and president of Young, Gifted and Black; Tricia Cortez of Laredo, Texas, Executive Director of the Rio Grande International Study Center; Vivian Moreno, of Chicago, Illinois and head of the Little Village Environmental Justice Organization; and Candice Youngblood of Los Angeles, a senior attorney at Earthjustice and founder of the EJ project Youth on Root.[178]

Despite becoming a national icon, it should be no surprise that Rachel Carson's legacy includes a love of particular places, communities, local action, and grassroots networks that are rarely counted or reported on sufficiently as a key part of the widespread and influential environment movement. As we have seen, Rachel fought hard to prevent the development of her beloved "Lost Woods" near her summer cottage in Southport, Maine. And she loved places like Nags Head and Beaufort on the Outer Banks of North Carolina. But she also was deeply engaged in the areas around the Maryland suburbs and Washington, D.C. where she spent most of her adulthood.

Rachel Carson was an active birder her whole life — from looking at and writing about her childhood friends like "Mr. and Mrs. Wren," to her time in Baltimore birding with and reviewing DDT research with Chan Robbins of the Patuxent Research Center, to joining the Audubon Naturalist Society (ANS) in 1947 when she first moved to Silver Spring, Maryland.[179] The ANS had been founded in 1897 by Florence Merriam Bailey, the sister of Teddy Roosevelt's friend, naturalist and author, C. Hart Merriam. Florence Merriam also was married to Roosevelt's friend and trusted field guide, Vernon Merriam. When Roosevelt became president, the Baileys moved to Washington as Hart Merriam became the head of the U.S. Biological Survey. A remarkable and immediate predecessor of Rachel Carson, Bailey had organized and written for the National Audubon Society following her years at Smith College where she was perhaps the first U.S. college campus environmental organizer. Bailey also wrote the first popular field guide to birds, introduced the use of binoculars to observe them, and wrote for many popular magazines including *St. Nicholas* which was read avidly by Rachel's mother, Maria Carson. Florence met with Roosevelt in the White House (he became a member of ANS which was then called the D.C. Audubon Society), wrote popular books about birds, taught classes across the area about birds and nature, and was part of the historic campaign to halt the slaughter of herons and egrets whose feathers were used to adorn fashionable women's hats. Although there is no record of Rachel Carson actually meeting Florence Merriam Bailey, she undoubtedly knew of her and would have been influenced by her. Bailey continued as the head of the D.C. Audubon Society until 1947 when she was well into her eighties.[180] Rachel Carson joined the Board of D.C. Audubon shortly after in 1948 with a new group of active members who revived the group which had declined since its early heyday. The new DC Audubon group included prominent birders, experts and authors like Roger Tory Peterson, Chan Robbins, Louis Halle, Irston Barnes, and Shirley Briggs who edited the society's newsletter, *The Thrush*, which included Carson's book re-

views. Much of Rachel's social activity revolved around birding with these local birders with national connections. They went to the Eastern Shores of Maryland and Virginia, camping and birding trips to Cobb Island off the Virginia Coast, and to a particular favorite of Rachel's, the towpath of the C&O Canal in Maryland which offers views of the Potomac River and spring warblers, vireos, Indigo Buntings, Scarlet and Summer Tanagers, Baltimore Orioles, kingfishers, herons, egrets, and more.[181]

Shirley Briggs and Diana Post who carried on Rachel's legacy as heads of the Rachel Carson Council also loved and walked along the C&O Canal in Rachel Carson's footsteps, as does the author, even decades before joining the Rachel Carson Council.[182] It is why each year the Rachel Carson Council brings together all of its future Rachel Carsons, its RCC Fellows, to its RCC American Environmental Leadership Institute (AELI), held on Capitol Hill and also at a retreat setting at Rockwood Manor that sits on a wooded site adjacent to the C&O canal where Rachel Carson walked, talked with friends and colleagues, birded, and sought inspiration and renewal.

At the AELI, the Rachel Carson Fellows explore the canal and Great Falls, go on nature walks, practice nature writing and journaling, and hear the flutelike songs of the wood thrushes and veerys that Rachel loved. And, when the AELI is held at the RCC Washington office and on Capitol Hill, like Rachel, they meet with Senators and members of Congress, environmental authors and leaders, and hone their skills in organizing, lobbying, and leadership. And, they form bonds of friendship and environmental networks that will serve to sustain them as they continue on and carry out the work of Rachel Carson and all those who came before and after her untimely death.

These young environmental leaders who embody the legacy of Rachel Carson are coming of age in dark, difficult times. But, then, so did Rachel who headed off excitedly in 1929 from Chatham University (then the Pennsylvania College for Women) to Johns Hopkins University and the Woods Hole Institute just as the Great Depression eclipsed her plans, her hopes, her dreams. But Rachel Carson was not alone even as she struggled to find a new path to use her gifts to protect the earth and those who share it. Surely those formative years of her youth heightened Rachel's desire to tell others about the awe and wonder of the natural world that sustained her and the need for each of us to act boldly to preserve a living environment that will sustain us all.

As her fame gave her the opportunity and as she faced death, Rachel Carson spoke of these things to a new generation. In her commencement address of 1962 to the women of Scripps College, Rachel urged the graduates to observe the world around them, to become involved, to take action, and to lead.[183]

Your generation must come to terms with the environment. Your generation must face realities instead of taking refuge in ignorance and evasion of truth. Yours is a grave and a sobering responsibility, but it is also a shining opportunity.
—Rachel Carson

Now, decades later, the environment movement that Rachel Carson reignited marches forward with ranks and ranks of organizations, with millions and millions of activists, with authors and scientists and academics and grassroots organizers and prophetic leaders who cry out for the environment and for justice and peace. In the iconic photograph of Rachel Carson seated in Olympian style atop the sandstone boulders of Hawk Mountain in Pennsylvania, with her stylish leather jacket and her binoculars, Carson's gaze is trained on the horizon of hills in the distance where hawks and eagles first glide into sight. But it is easy to imagine that Rachel is, instead, looking to the future. What she sees is you and me, along with lines of countless environmentalists, and the endless, eager faces of the new, young generation of environmental leaders, of Rachel Carsons, who will carry on her work, her legend and her legacy.

Endnotes

1. Rachel Carson grew up in Springdale, Pennsylvania in a small farmhouse that is now the site of the Rachel Carson Homestead, https://rachelcarsonhomestead.org/. For much of the time she worked in Washington and wrote *Silent Spring*, Carson lived at 11701 Berwick Road in Silver Spring, Maryland, the site of the Rachel Carson Landmark Alliance, https://rachelcarsonlandmarkalliance.org/. Her Maine summer cottage is on Southport Island, Maine, and is owned by her son, Roger Christie. See Bob Musil, "Rachel Carson's Cottage at the Edge of the Sea," https://rachelcarsoncouncil.org/wp-content/uploads/2019/11/RC_Cottage_Essay.pdf. The Rachel Carson National Wildlife Refuge (NWR) is headquartered in Wells, Maine and runs for over 50 miles along the Maine coast. https://www.fws.gov/refuge/rachel-carson.The Rachel Carson College is part of the University of California, Santa Cruz and is dedicated to sustainability and the principles of Rachel Carson. https://rachelcarson.ucsc.edu/. Rachel Carson is pictured on a U.S. 17¢ postage stamp issued in 1981. https://postalmuseum.si.edu/exhibition/women-on-stamps-part-2-science-improving-lives/rachel-carson. Among books on Rachel Carson for children and youth, one of the best is by Laurie Lawlor, *Rachel Carson and Her Book That Changed the World* (2012), available widely online. The Rachel Carson Bridge spans the Allegheny River in Pittsburgh. Formerly the Ninth Street Bridge, its name was changed on Earth Day 2006 after years of lobbying by Esther Barazzone, the President of Chatham University, Carson's alma mater. https://en.wikipedia.org/wiki/Rachel_Carson_Bridge. Carson is also honored with a memorial star, much like those in Hollywood, on the Pittsburgh Walk of Fame. https://pittsburghwalkoffame.org/rachel-carson/. Rachel Carson was inducted into the Irish-American Hall of Fame in 2020. https://iahof.org/inductees-2020/.

2. Linda Lear, *Rachel Carson: Witness for Nature*, (New York: Henry Holt and Company, Owl Books, 1998), p. 217. Also see Rachel Carson, "Remarks at the Acceptance of the National Book Award for Non-Fiction" in Linda Lear, ed., with an Introduction by Linda Lear, *Lost Woods: The Discovered Writing of Rachel Carson* (Boston: Beacon Press, 1998, 90-92.

3. Rachel Carson, *Silent Spring*, (Boston: Houghton Mifflin Company, 1962).

4. As Rachel Carson was writing *Under the Sea-Wind*, advanced Technicolor with full color and sound first reached large audiences through feature

films like *Snow White and the Seven Dwarfs* in 1937 and *The Wizard of Oz* and *Gone with the Wind* in 1939. Walt Disney had first produced an animated nature film in three colors with an earlier, simpler version of Technicolor in 1932, but it was the first in his "Silly Symphony series," *Flowers and Trees*. William Beebe provided photos from his Bathysphere in the *National Geographic* in 1931, but sea creatures were presented through paintings by the artist Else Bostelmann. His *Half Mile Down* (1934) included eight color plates, but again, sea creatures, such as one of an Orange-lighted Finger Squid was painted by Bostelmann. The French filmmaker, Jean Painlévé, had made a short black-and-white motion picture of a crawling octopus in 1931, though it was not in the wild. Henry Bigelow's *Fishes of the Gulf of Maine* (1925) that Carson used as a primary source contained only black-and-white photos and line-drawn illustrations.

5. Susan Power Bratton, "Thinking Like a Mackerel: Rachel Carson's *Under the Sea-Wind* as a Source for a Transformational Sea Ethic," *Ethics and the Environment* (Spring, 2004, 9, 1), 1-22. Also, slightly edited, in Lisa H. Sideris and Kathleen Dean Moore, eds. *Rachel Carson: Legacy and Challenge* (Albany: State University of New York Press, 2008), 79-93. Also, see, Sandra Gibson, "Rachel Carson: The Non-human Perspective," *Nerve*, 28, July 16, 2020 at: http://www.catalystmedia.org.uk/nerve/rachel-carson-the-non-human-perspective/. Gibson cites Carson's use of eels in *Under the Sea-Wind* as evidence of her empathy.

6. Rachel Carson, "The Real World Around Us," speech to Theta Sigma Pi, The Sorority of Women Journalists, April 21, 1954. Rachel Carson Papers, Beinecke Library of Rare Books and Manuscripts, Yale University. Also, in Lear, *Lost Woods*, 149. Also see, Lear, *Witness for Nature*, 80.

7. From Rachel Carson's acceptance speech for the National Book Award for Nonfiction, January 27, 1952, Rachel Carson Papers, Beinecke Library. In Lear, *Lost Woods*, 91. Also, Lear, *Witness for Nature*, 219.

8. "The World of Waters" MSS, 1, 1936, Rachel Carson Papers, Beinecke Library. There are a number of drafts of "The World of Waters," Carson's attempt at the report for Elmer Higgins of the Bureau of Fisheries, including a last revision at the request of the *Atlantic* that was published as "Undersea" in September, 1937.

9. Lear, *Witness for Nature*, 19, 33-34.

10. Lear, *Witness for Nature*, 79, 82-83.

11 Lear, *Witness for Nature*, 70-74.

12. Lear, *Witness for Nature*, 78-79.

13. Lear, *Witness for Nature*, 93-94. Also see Robert K. Musil, "What Might Have Been," Rachel Carson Council, https://rachelcarsoncouncil.org/about-rcc/about-rachel-carson/rachel-carsonwhat-might-have-been/.

14. Carson, "The Real World Around Us," Rachel Carson Papers, Beinecke Rare Book and Manuscript Library, Yale University. Also, in Lear, *Lost Woods*, 150. See, too, Lear, *Witness for Nature*, 81.

15. R.L. Carson, "Undersea" *The Atlantic Monthly*, 160, no. 3 (September, 1937), 322-325. Also, in Lear, *Lost Woods*, 3-11. See, too, Lear, *Witness for Nature*, 81-82, 85-87.

16. Quincy Howe to Hendrik van Loon, September 9, 1937 and Hendrik van Loon to Rachel Carson, September 10, 1937, Rachel Carson Papers, Beinecke Library. Rachel Carson, "The Real World Around Us." Also see, Lear, *Witness for Nature*, 88-90.

17. Rachel Carson, *Under the Sea-Wind*, Introduction by Linda Lear, Centennial Edition (New York: Penguin Books, 2007), 9.

18. Paul H. Oehser, "Living Things," *The Washington Post*, December 7, 1941, L10.

19. William Beebe, "Of and About the Sea," *The Saturday Review of Literature*, December 27, 1941, 5.

20. Rachel L. Carson, "The Odyssey of the Eel," in William Beebe, *The Book of Naturalists*, (New York: Alfred A. Knopf, 1944), 478-495.

21. Carson, "The Real World Around Us." Also, Lear, *Witness for Nature*, 105, 110-115.

22. Rachel L. Carson, *The Sea Around Us* (New York, Oxford University Press, 1951) and *The Edge of the Sea* (Boston, Houghton Mifflin Company, 1955).

23. Lewis Herber (pseudonym for Murray Bookchin), *Our Synthetic Environment* (New York, Alfred A. Knopf, 1962). Bookchin's first book was credible and perceptive, but was written from a leftist perspective that failed to reach a large audience. He went on to become an influential writer for the New Left and Fellow at the Institute for Policy Studies, promoting a

number of decentralized, local sustainability projects.

24. For the importance of Roger Revelle's ocean expeditions, see D. James Baker and Walter H. Munk, "Roger Revelle, Oceanographer, 1909-1991," *Oceanography*, Vol. 4, No. 2, 1991. https://tos.org/oceanography/assets/docs/4-2_baker1.pdf. Also see, Walter H. Munk, "Tribute to Roger Revelle and his contributions to studies of carbon dioxide and climate change," Proceedings of the National Academy of Science, USA1997, Aug5; 94(16):8275-8279 at: https://pmc.ncbi.nlm.nih.gov/articles/PMC33716/.

25. Sir John Murray and Johan Hjort, *The Depths of the Ocean: A General Account of the Modern Science of Oceanography Based Largely on the Scientific Researches of the Norwegian Steamer Michael Sars in the North Atlantic* (London, Macmillan and Company, 1912). Rachel Carson relied heavily on the *Depths of the Ocean* and other studies she encountered at the Woods Hole Biological Laboratory.

26. The Official William Beebe Website, "Bathysphere," at: https://sites.google.com/site/cwilliambeebe/Home/bathysphere and William Beebe, *Half Mile Down* (New York: Harcourt Brace and Company, 1934).

27. William Beebe, "A Round Trip to Davy Jones's Locker," *National Geographic*, June 1931, 677.

28. Rachel Carson, "Memo to Mrs. Eales on *Under the Sea-Wind*," Rachel Carson Papers, Beinecke Library. Also in Lear, *Lost Woods*, 53-62.

29. Carson, *Under the Sea-Wind*, 3.

30. Carson, "Memo to Mrs. Eales," Lear, *Lost Woods*, 54.

31. Carson, "Memo to Mrs. Eales," Lear, *Lost Woods*, 54.

32. Carson, "Memo to Mrs. Eales," Lear *Lost Woods*, 56.

33. Carson, "Memo to Mrs. Eales," Lear, *Lost Woods*, 56.

34. See generally, Jonathan Balcombe, *What a Fish Knows: The Inner Lives of Our Underwater Cousins* (New York, Scientific American/Farrar, Straus & Giroux, 2016). Also see, S. H. Munsch, J. R. Cordell, and J. D. Toft, "Fine-Scale Habitat Use and Behavior of a Nearshore Fish Community: Nursery Functions, Predation Avoidance, and Spatiotemporal Habitat Partitioning," *Marine Ecology Progress Series* 557 (2016): 1–15, DOI: 10.3354/meps11862.

35. Carl Safina, *Beyond Words: How Animals Think and Feel* (New York: Hen-

ry Holt & Company, 2015) and *Eye of the Albatross: Visions of Hope and Survival* (New York: Henry Holt & Company, 2001); Jennifer Ackerman, *The Genius of Birds* (New York: Penguin Press, 2016) and *The Bird Way: A New Look at How Birds Talk, Work, Play, Parent, and Think* (Penguin Press, 2020); Deborah Cramer, *The Narrow Edge: A Tiny Bird, an Ancient Crab & an Epic Journey* (New Haven: Yale University Press, 2015) and *Great Waters: An Atlantic Passage* (New York: W.W. Norton & Company, 2001).

36. Carson, *Under the Sea-Wind*, 11.

37. K. von Frisch, "The Sense of Hearing in Fish" *Nature*, 141 (Jan. 1, 1938) 8-11. Also see Balcombe, *What a Fish Knows*, 41-43.

38. For the belief among many credible writers before the 1930s that fish could hear, see G.H. Parker, "A Critical Survey of the Sense of Hearing in Fishes," *Proceedings of the American Philosophical Society* 57, no. 2 (1918), 69-98 and Andrew Lang, *Introduction to The Compleat Angler*, Gutenberg e-Books at: https://www.gutenberg.org/files/2422/2422-h/2422-h.htm.

39. Carson, *Under the Sea-Wind*, 18.

40. Carson, *Under the Sea-Wind*, 16.

41. Carson, *Under the Sea-Wind*, 19.

42. Carson, *Under the Sea-Wind*, 78-79.

43. Carson, *Under the Sea-Wind*, 81-82.

44. Carson, *Under the Sea-Wind*, 93.

45. Carson, *Under the Sea-Wind*, 121.

46. Carson, *Under the Sea-Wind*, 116.

47. Rachel Carson, letters to Mary Frye, March 14 and April 24, 1928, Rachel Carson Papers, Beinecke Library, show Carson's interest in eels in graduate school. Also see Lear, *Witness for Nature*, 75 and 75 n58. Carson later wrote a newspaper piece about them, Rachel Carson, "Chesapeake Eels Seek the Sargasso Sea" *Baltimore Sunday Sun*, October 9, 1938. Also, in Lear, *Lost Woods*, 19-23. See, too, William Souder, *On a Farther Shore: The Life and Legacy of Rachel Carson* (New York, Crown Publishers, 2012), 53-54.

48. For eels sense of smell, see C.L. Deelder, "Synopsis of Biological Data on the Eel, Anguilla anguilla (Linnaeus, 1758), *FAO Fisheries Synopsis*, 80, Revision 1, 1970, 14-15 at: http://www.fao.org/3/a-ap945e.pdf. For

mapping eels, see Jason Bittel, "Epic Eel Migration Mapped for the First Time," *National Geographic,* October 27, 2015 at: https://www.nationalgeographic.com/news/2015/10/151027-american-eel-migration-animal-behavior-oceans-science/; For eels magnetic detection ability, see Lewis C. Naisbett-Jones, et. al., "A Magnetic Map Leads Juvenile European Eels to the Gulf Stream, *Current Biology*, 27, issue 8, 1236-1240, DOI: 10.1016/j.cub.2017.03.015; For an excellent summary of eel status and behavior, see Deborah Cramer, *Great Waters: An Atlantic Passage* (New York, London: W.W. Norton & Company, 2001), 192-193.

49. Carson, *Under the Sea-Wind*, 130.

50. Carson, *Under the Sea-Wind*, 145. Octopuses are the subject of numerous books about their behavior and intelligence. See Jennifer A. Mather, Roland C. Anderson and James B. Wood, *Octopus: The Ocean's Intelligent Invertebrate: A Natural History* (Portland, OR: The Timber Press, 2010); Sy Montgomery, *The Soul of an Octopus* (New York: Atria Books, 2015) and Peter Godfrey-Smith, *Other Minds: The Octopus, the Sea, and the Deep Origins of Consciousness* (New York: Farrar, Straus and Giroux, 2016). Also see the Netflix film, *My Octopus Teacher* at https://www.netflix.com/title/81045007. No records of the shipwreck of the *Mary B.* have been found. The scene is set above the Capes of Virginia, an area Carson studied and visited while living at Stemmers Run, 10 miles northeast of Baltimore where she worked for the Bureau of Fisheries. Carson describes fossilized sharks' teeth found around the Calvert Cliffs of Maryland, now a state park: https://dnr.maryland.gov/publiclands/pages/southern/calvertcliffs.aspx.

51. Carson, *Under the Sea-Wind*, 146,148-149.

52. Carson, *Under the Sea-Wind*, 149.

53. Carson, *Under the Sea-Wind*, 151.

54. Carson, *Under the Sea-Wind*, 151.

55. Carson, *Under the Sea-Wind*, 10.

56. Carson, *Under the Sea-Wind*, 11.

57. Carson, *Under the Sea-Wind*, 14-15. *The Baltimore Sun* artist, Howard Frech, whom Carson chose to produce the illustrations for the first edition of *Under the Sea-Wind* drew this dramatic moment of multiple predation and survival. It was included immediately before the Foreword in the 2007 centennial edition by Linda Lear.

58. Carson, *Under the Sea-Wind*, 15.

59. Rachel Carson, "Our Ever-Changing Shore" *Holiday*, 24 (July, 1958), 71-120. Also, in Lear, *Lost Woods*, 114. See Lear, *Witness for Nature*, 306-309.

60. Arthur Cleveland Bent, *Life Histories of American Shore Birds, Part 1* (New York, Dover Publications, 1962), 265-277. The Dover edition is a republication of Smithsonian Institution United States National Museum *Bulletin 142* that Carson would have used.

61. Carson, *Under the Sea-Wind*, 42, 44.

62. Carson, *Under the Sea-Wind*, 81-82.

63. Carson, *Under the Sea-Wind*, 86.

64. Carson, *Under the Sea-Wind*, 154-155.

65. Carson, "Memo to Mrs. Eales," Lear, *Lost Woods*, 57; Sam Bland, "A Sign of Fall: Mullet Fishing on Bogue Banks," *Coastal Review Online*, September 23, 2014, https://www.coastalreview.org/2014/09/a-sign-of-fall-mullet-fishing-on-bogue-banks/ for the history and culture of fishing the mullet run in North Carolina. For the continuing contemporary gathering of fishermen, see Nick Honachefsky, "East Coast Fall Mullet Run," Salt Water Sportsman Online (Salt Water Sportsman, August 10, 2018), https://www.saltwatersportsman.com/east-coast-fall-mullet-run/.

66. Carson, *Under the Sea-Wind*, 59.

67. Carson, *Under the Sea-Wind*, 63.

68. *New York Times*, "Two Best Sellers," April 27, 1952, 163. Paul Brooks, *The House of Life*, 130. Lear, *Witness for Nature*, 226.

69. Sterling North, "Miss Carson Dramatizes Sea Battle to Survive," *The Washington Post*, April 6, 1952, B6.

70. *New York Times*, "Two Best Sellers," 163.

71. Henry Beston, *A Year of Life on the Great Beach of Cape Code*, (New York: Henry Holt & Company, 1949). For Carson's admiration of Henry Beston, with whom she eventually became friends, see Lear, *Witness for Nature*, 101-102, 238.

72. Henry Beston, "Miss Carson's First," *The Freeman*, 3, no. 3 (November 3,

1952), 100.

73. For the unprecedented response to *The Sea Around Us* that was praised by academics, scientists, the public and set off a tsunami of adulation for Carson, see Lear, *Witness for Nature*, 199-210. It was on the *New York Times* best-seller list for eighty-six weeks. *Under the Sea-Wind* was published in the shadow of this acclaim, yet was deemed "a classic" and compared to literary giants. *The Edge of the Sea* completed a trifecta of awards, esteem, and sales for Carson. The *Times* summed it up that Miss Carson "has done it again." Lear, *Witness for Nature*, 226, 275. Also see, Souder, *On a Farther Shore*, 152-157.

74. Rachel Carson, *Silent Spring* (Boston, Houghton Mifflin Company, 1962), 25-26, 30, 33.

75. Carson, *Silent Spring*, 106-109.

76. Lear, *Witness for Nature*, 118-119,118n27, Robert K. Musil, "Ode to Federal Bureaucrats," Rachel Carson Council, https://rachelcarsoncouncil.org/ode-federal-bureaucrats/.

77. Lear, *Witness for Nature*, 118-119, 119n28,119n29, Robert K. Musil, "What Might Have Been," Rachel Carson Council, https://rachelcarsoncouncil.org/about-rcc/about-rachel-carson/rachel-carsonwhat-might-have-been/.

78. Paul Brooks, *The House of Life: Rachel Carson at Work* (Boston: Houghton Mifflin, 1972), 6-7.

79. Brooks, *House of Life*, 7-8.

80. Brooks, *House of Life*, 35.

81. Henry Williamson, *Salar the Salmon*, (London: Faber and Faber Limited, 1935), 15-16. Also see Lear, *Witness for Nature*, 90 and 90n29 and Souder, *On a Farther Shore*, 59-63.

82. Henry Williamson, with an Introduction by Sir John Fortescue, *Takar the Otter: His Joyful Water-Life and Death in the Country of the Two Rivers* (New York: E.P. Dutton & Company, 1928). Roger Christie on being read *Takar*, personal recollection with author, July, 2020.

83. Rachel Carson, letter to Dorothy Freeman, November 20, 1953 in Martha Freeman, ed., *Always, Rachel: The Letters of Rachel Carson and Dorothy*

Freeman, 1952-1964 (Boston: Beacon Press, 1995), 11.

84. Lear, *Witness for Nature*, 39-40, Alfred, Lord Tennyson, "Locksley Hall," Poetry Foundation, https://www.poetryfoundation.org/poems/45362/locksley-hall.

85. Tennyson, "Locksley Hall."

86. Souder, *On a Farther Shore*, 77.

87. Tennyson, "Locksley Hall."

88. Algernon Charles Swinburne, "A Forsaken Garden," Poetry Foundation, https://www.poetryfoundation.org/poems/45287/a-forsaken-garden.

89. Swinburne, "A Forsaken Garden."

90. Rachel Carson, "Foreword," *Under the Sea-Wind*, 3.

92. Lear, *Witness for Nature*, 389.

93. John Elder, "Withered Sedge and Yellowed Wood: Poetry in Silent Spring," in Peter Matthiessen, ed., *Courage for the Earth: Writers, Scientists, and Activists Celebrate the Life and Writings of Rachel Carson* (Boston and New York: Houghton Mifflin Company, 2007), 80-81.

94. Lear, *Witness for Nature*, 13-14. Also see Robert K. Musil, *Rachel Carson and Her Sisters: Extraordinary Women Who Have Shaped America's Environment* (New Brunswick, NJ and London: Rutgers University Press), 12-13.

95. Musil, *Rachel Carson and Her Sisters*, 32-33, 42.

96. Lear, *Witness for Nature*, 16-17; Musil, *Rachel Carson and Her Sisters*, 12.

97. Lear, *Witness for Nature*,18-20; Musil, *Rachel Carson and Her Sisters*, 91, 93.

98. Linda Lear, "The Next Page: Rachel Carson, a Child of the Allegheny," *Pittsburgh Post-Gazette*, July 8, 2018 at: https://www.post-gazette.com/opinion/Op-Ed/2018/07/08/The-Next-Page-Child-of-the-Allegheny/stories/201807080009.

99. Lear, *Lost Woods*, 32. Today, ocean acidification caused by climate change is beginning to dissolve shells made from calcium carbonate like the tiny ones of foraminifera which over the eons have formed limestone formations like those at Hawk Mountain. See, Earle, *The World is Blue*, 178-180.

Also S. Uthicke, et. al., "High Risk of Extinction of Benthic Foraminifera in This Century, *Nature*, Scientific Reports 3, no.1 (2013): 1769 at: https://www.nature.com/articles/srep01769#citeas; and Fabricio Guamán-Guevara, et. al., "Impacts of Ocean Acidification on Intertidal Benthic Foraminiferal Growth and Calcification," *PLOS One*, (2019), DOI: 10.1371/journal.pone.0220046.

100. Lear, *Witness for Nature*, 50-52, 69-70, 72-74.

101. Lear, *Witness for Nature*, 428-430.

102. Sir John Murray and Johan Hjort, *The Depths of the Ocean* (London: Macmillan, 1912), https://news.northwestern.edu/stories/2020/08/provost-kathleen-hagerty/?utm_medium=email&utm_source=iModules%20email&utm_campaign=2020-08-15%20Purple%20Line%20domestic; and Henry B. Bigelow and William W. Welsh, *Fishes of the Gulf of Maine* (Washington, DC: Government Printing Office, 1925) is available at the U.S. Fish and Wildlife Conservation Library, Shepherdstown, West Virginia. The digitized, online 1925 edition is at BHL: The Biological Heritage Library: https://www.biodiversitylibrary.org/item/29916#page/1/mode/1up. Carson read Bigelow's many articles on sea life in addition to *Fishes of the Gulf of Maine.* She greatly admired him, became his friend, and sought his advice. A visit with him encouraged her to write *The Sea Around Us* which she dedicated to Bigelow in its revised tenth anniversary edition in 1961. Lear, *Witness for Nature*, 139, 180. Also see Martha Freeman, ed., *Always, Rachel: The Letters of Rachel Carson and Dorothy Freeman, 1952-1964* (Boston: Beacon Press, 1995), 12, 35, 351.

103. See Rachel L. Carson, "Chesapeake Eels Seek the Sargasso Sea" *Baltimore Sunday Sun*, October 9, 1938 in Lear, *Lost Woods*, 19-23. Also see Lear, *Witness for Nature*, 82-83.

104. *CBS Reports*, "The Silent Spring of Rachel Carson," April 3, 1963, available on CD from the Rachel Carson Council at: https://rachelcarsoncouncil.org/product/the-silent-spring-of-rachel-carson/. For PBS American Experience, "Rachel Carson's Silent Spring," 1993, see Rachel Carson Center for Environment and Society, Environment and Society Portal, "Rachel Carson on Television," http://www.environmentandsociety.org/exhibitions/rachel-carsons-silent-spring/silent-spring-television#:~:text=On%203%20April%201963%2C%20CBS,Stevens%2C%20spokesman%20for%20the%20agricultural. The 1993

PBS American Experience, "Rachel Carson is available at: https://shop.pbs.org/WA9802.html. Only in 2018 did PBS revisit Rachel Carson with a fresh perspective that included serious discussion of her ocean works and compelling footage of the sea. A new documentary on the PBS American Experience, "Rachel Carson," is available at: https://www.pbs.org/wgbh/americanexperience/films/rachel-carson/. For a review that includes observations on Carson's life and ocean works, as well as *Silent Spring*, see "Film Review: PBS American Experience's 'Rachel Carson,'" Earthzine, February 8, 2020, https://earthzine.org/film-review-pbs-american-experiences-rachel-carson/.

105. Sandra Steingraber, "*Silent Spring*: A Father-Daughter Dance" in Matthiessen, *Courage for the Earth,* 56.

106. Rachel Carson, *Silent Spring,* with an Introduction by Vice President Al Gore (Boston, Houghton Mifflin, 1994).

107. Linda Lear, *Witness for Nature.*

108. Lear, *Lost Woods*, 3-11, 19-23, 53-62.

109. Linda Lear, "Introduction", Rachel Carson, *Under the Sea-Wind,* centennial edition, with an Introduction by Linda Lear (New York: Penguin Classics, 2007). ix-xx and Linda Lear, "Remembrance of Life," in Matthiessen, *Courage for the Earth*, 183-196. For Lear's assessment of *Under the Sea-Wind,* see 186.

110. Janisse Ray, "Changing Sex," 109-128 in Matthiessen, *Courage for the Earth*, 109-128. For other women who have been influenced by Carson and carried her work forward as writers and advocates, see Musil, *Rachel Carson and Her Sisters.*

111. Mark Hamilton Lytle, *The Gentle Subversive: Rachel Carson, Silent Spring, and the Rise of the Environmental Movement* (Oxford, New York: Oxford University Press, 2007), 223-226. For an accurate description of the treaty and DDT for which right-wing critics attempt to smear Carson and the environment movement, see Musil, *Rachel Carson and Her Sisters*, 186-187.

112. Lisa H. Sideris and Kathleen Dean Moore, eds., *Rachel Carson: Legacy and Challenge* (Albany: State University of New York Press, 2008), 1.

113. Jill Lepore, "The Right Way to Remember Rachel Carson," *The New Yorker*, March 19, 2018; Sandra Steingraber, ed., *Silent Spring & Other Writings on the Environment* (New York: The Library of America, 2018). A Ph.D. sci-

entist, Steingraber's Introduction is brilliant in explaining how advanced and accessible Carson's science writing in *Silent Spring* really is. To be fair, her volume seeks to trace the development of Carson's advocacy which is not directly on display in the ocean works. Steingraber did note, however, that the ocean books "are credited by contemporary ecologists for having anticipated complexity theory." At the time, she was also planning to edit an ocean volume for the LOA, now published, and has written admiringly about all of Carson's writings. See Sandra Steingraber, ed., *Rachel Carson: The Sea Trilogy: Under the Sea-Wind, The Sea Around Us, The Edge of the Sea* (New York: The Library of America, 2021).

114. Lear, "The Next Page: Rachel Carson, a Child of the Allegheny," *Pittsburgh Post-Gazette*, July 8, 2018.

115. Rachel Carson, *The Sea Around Us*, Introduction by Sylvia Earle (New York: Oxford University Press, 2018), xi.

116. Rachel L. Carson, *The Sea Around Us*, Introduction by Ann H. Zwinger, Afterword by Jeffrey S. Levinton, special edition (New York and Oxford: Oxford University Press, 1961). In her new Preface, Carson wrote in detail with cutting-edge science about the impact of nuclear testing and nuclear wastes. Also, in Lear, *Lost Woods*, 101-109.

117. Deborah Cramer, *Great Waters: An Atlantic Passage* (New York and London: W.W. Norton & Company, 2001), 28; Cramer, *Smithsonian Ocean: Our Water, Our World* (New York: Smithsonian Institution/Collins), 2008; Sylvia A. Earle, *Blue Hope: Exploring and Caring for Earth's Magnificent Ocean* (Washington, DC: National Geographic, 2014), 75-76.

118. Andrew Freedman, "Hottest Arctic Temperature Record Probably Set with 100-degree Reading in Siberia," *The Washington Post Online* (*The Washington Post*, June 23, 2020), https://www.washingtonpost.com/weather/2020/06/21/arctic-temperature-record-siberia/.

119. Roger Revelle, Charles Keating and others issued a report through the Conservation Foundation in 1963 on the findings of a March 12, 1963 climate change conference that warned of rising global temperatures if CO2 emissions were not curbed. Conservation Foundation, *Implications of the Rising Carbon Dioxide Content of the Atmosphere*, 1963, https://babel.hathitrust.org/cgi/pt?id=mdp.39015004619030&view=1up&seq=1. Rachel Carson was friends with Revelle and regularly read Conservation Foundation Reports. Also see, Bob Musil, "The Tragedy of Climate

Change Reports," *Huffington Post*, January 20, 2015, https://www.huffpost.com/entry/the-tragedy-of-climate-ch_b_6502804?guccounter=1&guce_referrer=aHR0cHM6Ly93d3cuZ29vZ2xlLmNvbS8&guce_referrer_sig=AQAAAIj48n8HWxB5lITIJ5irh13-qltf4WifOJ63J16k2bDKXEv2N2t-RMxVGs655MZHraWPkT6kcq_H_4MUN0OwN2j4-7Rwe0-bYIbYc0k-w5zT_ZmJKAcaMBV0lwUomCxvo7RfP1hTFfntM13mlnjhSNQMreUm-tAEMtdX_Zi9klbU1D. See also Souder, *On a Farther Shore*, 372-373.

120. Doyle, Rice, "Earth's Carbon Dioxide Levels Are Highest They Have Been in Millions of Years," *USA Today*, June 4, 2019 at: https://www.usatoday.com/story/news/nation/2019/06/04/carbon-dioxide-levels-hawaiis-mauna-loa-soar-record-peak/1337163001/. For 2021 climate data see Zeke Hausfather, "State of the Climate: How the World Warmed in 2021," Carbon Brief, January 17, 2022, https://www.carbonbrief.org/state-of-the-climate-how-the-world-warmed-in-2021. For initial and current readings at Mauna Loa see Scripps Institution of Oceanography, "The Keeling Curve," at: https://scripps.ucsd.edu/programs/keelingcurve/. For more on how the ocean impacts carbon emissions, see Oregon State University, "Microbiology research furthers understanding of ocean's role in carbon cycling," *ScienceDaily*, October 7, 2021, www.sciencedaily.com/releases/2021/10/211007170631.htm.

121. "Climate Change in the Artic," National Snow & Ice Data Center, Last updated May 4, 2020, https://nsidc.org/cryosphere/arctic-meteorology/climate_change.html#:~:text=Over%20the%20past%2030%20years,of%20human%2Dcaused%20climate%20change. For more on the destabilizing effect of climate change in the Arctic, see Sarah Kaplan, "Climate Change Has Destabilized the Earth's Poles, Putting the Rest of the Planet in Peril," *The Washington Post* (WP Company, December 14, 2021), https://www.washingtonpost.com/climate-environment/2021/12/14/climate-change-arctic-antarctic-poles/.

122. "Ocean Acidification," The Ocean Portal (The Smithsonian Institute, April 2018), https://ocean.si.edu/ocean-life/invertebrates/ocean-acidification.

123. For overall sea-level rise, see "Is Sea Level Rising? Yes, Sea Level Is Rising at an Increasing Rate," National Ocean Service (National Oceanic and Atmospheric Association), last modified February 22, 2021, https://oceanservice.noaa.gov/facts/sealevel.html and Tony Greicius, "Study Projects a Surge in Coastal Flooding, Starting in 2030s," NASA.gov (NASA, July 7, 2021), https://www.nasa.gov/feature/jpl/study-projects-a-surge-in-coastal-

flooding-starting-in-2030s. For recent rise on the coast in *Under the Sea-Wind*, see Jim Morrison, "Flooding Hot Spots: Why Seas Are Rising Faster on the U.S. East Coast," Yale E360 (Yale School of the Environment, April 24, 2018), https://e360.yale.edu/features/flooding-hot-spots-why-seas-are-rising-faster-on-the-u.s.-east-coast.

124. Flemming T. Dahlke, Sylke Wohlrab, Martin Butzin, and Hans-Otto Pörtner, "Thermal bottlenecks in the life cycle define climate vulnerability of fish," *Science* 369, no. 6499, (2020), 65-70, DOI: 10.1126/science.aaz3658; Adam Vaughan, "Climate Change Will Make World Too Hot for 60 per cent of Fish Species," *New Scientist*, July 2, 2020 at: https://www.newscientist.com/article/2247602-climate-change-will-make-world-too-hot-for-60-per-cent-of-fish-species/?ct=t(RSS_EMAIL_CAMPAIGN).

125. https://www.google.com/url?sa=t&source=web&rct=j&opi=89978449&url=https://assets.wwf.org.uk/downloads/living_blue_planet_report_2015.pdf&ved=2ahUKEwi6jufq37mUAxXnK1kFHcejJYoQFnoECBoQAQ&usg=AOvVaw24DWg1C3N_JvQxlKW007_U.

126. Karla B. Parga Martinez, Mine B. Tekman, and Melanie Bergmann, "Temporal Trends in Marine Litter at Three Stations at the Hausgarten Observatory in the Arctic Deep Sea," *Frontiers in Marine Science*, 7, (2020), https://www.frontiersin.org/articles/10.3389/fmars.2020.00321/full.

127. For hermit crabs, see, Nicola Davis, "Microplastics Disrupt Hermit Crabs' Ability to Choose Shell, Study Suggests," The Guardian Online (The Guardian, April 29, 2020), https://www.theguardian.com/environment/2020/apr/29/microplastics-disrupt-hermit-crabs-ability-to-choose-shell-study-suggests. Hermit crabs are in *Under the Sea-Wind*, 15. Ironically, fiddler crabs were also harmed by DDT when it was sprayed over salt marshes before it was banned. See John Hay, "A Long View of Rachel Carson," in Matthiessen, *Courage for the Earth*, 104.

128. For mercury levels in the upper ocean, see Anne Casselbaum, "Humans Have Tripled Mercury Levels in Upper Ocean," (Nature, 2014), https://www.nature.com/news/humans-have-tripled-mercury-levels-in-upper-ocean-1.15680.

129. Alice Lipscombe-Southwell, "Toxic Mercury Pollution Found in the Ocean's Deepest Point," BBC Science Focus Magazine (BBC, June 28, 2020), https://www.sciencefocus.com/news/toxic-mercury-pollution-found-in-the-oceans-deepest-point/#:~:text=Now%2C%20two%20inde-

pendent%20teams%20of,%2C%20from%20Tianjin%20University%2C%2-0China.

130. Cramer, *Great Waters: An Atlantic Passage*, 199-200.

131. Earle, *The World is Blue*, 53-55.

132. For Audubon describing endless passenger pigeons in 1813, see https://manliuslibrary.org/double-elephant-folio/59-passenger-pigeon/. Martha, the last surviving passenger pigeon, died in the Cincinnati Zoo on September 1, 1914. She is now on display at the Smithsonian National Museum of Natural History in the "Objects of Wonder" exhibition. https://naturalhistory.si.edu/research/vertebrate-zoology/birds/collections-overview/martha-last-passenger-pigeon.

133. Rachel Carson, "Mr. Day's Dismissal," in Lear, *Lost Woods*, 98-100.

134. Rachel Carson, "National Women's Press Club Speech," in Lear, *Lost Woods*, 201-210.

135. Rachel Carson, "Foreword" to Ruth Harrison, *Animal Machines: The New Factory Farming Industry* (London: Vincent Stuart, LTC., 1964). Also, in Lear, *Lost Woods*, 194-196.

136. Souder, *On a Farther Shore*, 77.

137. Douglas Brinkley, "Rachel Carson and JFK, an Environmental Tag Team," *Audubon*, May-June 2012. Also see Lear, *Witness for Nature*, 375-376.

138. Edward O. Wilson, "On Silent Spring," in Matthiessen, *Courage for the Earth*, 33.

139. Lear, *Witness for Nature*, 483 and 483n 85. For the contemporary Rachel Carson Council, see www.rachelcarsoncouncil.org. The archives of the Rachel Carson Council are at the National Conservation Training Center, U.S. Fish and Wildlife Service, Shepherdstown, West Virginia.

140. Lear, *Witness*, 483, 583 fn85. Originally named the Rachel Carson Living Trust for the Environment, the organization was renamed in 1978 as the Rachel Carson Council.

141. Lear, *Witness*, 483.

142. 142. Lear, Witness, 181-182. The Audubon Naturalist Society changed its name to Nature Forward in October, 2022 https://natureforward.org/our-commitment-to-each-other/. It is located at the Woodend Sanctuary and Mansion in Chevy Chase, Maryland where the Rachel Carson Council was

once housed. Nature Forward (ANS) archives, including *The Thrush* and *The Audubon Naturalist*, are at the Smithsonian Institution.

143. Shirley A. Briggs, *Basic Guide to Pesticides: Their Characteristics and Hazards*, (Washington, D.C. Taylor & Francis, 1992). There are several different printings of *Basic Guide* which was published by the Rachel Carson Council through small publishing houses.

144. The Rachel Carson Landmark Alliance https://rachelcarsonlandmarkalliance.org/ and the Rachel Carson Landmark House are at 11701 Berwick, Silver Spring, Maryland. Linda Lear maintains a website focused on Rachel Carson https://www.rachelcarson.org/ and a Linda Lear Center for Special Collections and Archives at Connecticut College with Rachel Carson materials from Lear's research and writing about her. https://www.conncoll.edu/library-information-technology//libraries/libraries-locations/linda-lear-center-for-special-collections-and-arc/.

145. See the Rachel Carson Council, "About President & CEO, Robert K. Musil, Ph.D., M.P.H. https://rachelcarsoncouncil.org/about-rcc/about-president-robert-k-musil/.

146. The RCC's 2016 national conference and celebration, "A Sense of Wonder: The National Rachel Carson Jubilee and Civic Colloquium" is available in video at https://www.youtube.com/watch?v=LAivzyowdsU&list=PL22zOAZiQJkQx30EHf9JtFbMv-IBS6kiB.

147. The Rachel Carson Council Campus Network is at https://rachelcarsoncouncil.org/campus-program/.

148. Rachel Carson Council Fellowship Programs and the writing and work of RCC Fellows are at https://rachelcarsoncouncil.org/rachel-carson-council-fellowship-programs/ and https://rachelcarsoncouncil.org/student-essays/. Duke University also has Rachel Carson Scholars in Marine Science.

149. The RCC's report *Pork and Pollution* is at https://rachelcarsoncouncil.org/pork-and-pollution/. Carson's Introduction to *Animal Machines* is in Ruth Harrison, *Animal Machines*, Introduction by Rachel Carson (London: Vincent Stuart Publishers, 1964). It is available through the Internet Archives https://archive.org/details/animalmachines0000harr/page/n231/mode/2up. For the complete Preface see, Linda Lear, ed., with an Introduction by Linda Lear, *Lost Woods: The Discovered Writing of Rachel Carson*, (Boston: Beacon Press, 1998), 194-196. Carson was also friends with Christine Stevens, the American animal rights activist and founder of the Animal

Welfare Institute (AWI) https://awionline.org/, where she served on the Board of Directors. See Lear, *Lost Woods,* 192-193 that includes Carson's "To Understand Biology," an introduction to an AWI booklet.

150 RCC's report *Fowl Matters* is at https://rachelcarsoncouncil.org/fowl-matters/.

151. Carson, *Silent Spring*, 26, 30, 33, 36.

152. Rachel Carson, "The Pollution of Our Environment," in Lear, *Lost Woods*, 237-238.

153. See a discussion of Marie Rodell's role as King's agent at the Stanford University King Institute https://kinginstitute.stanford.edu/rodell-marie-freid#:~:text=Marie%20F.%20Rodell%20served%20as%20Martin%20Luther,of%20his%20first%20book%2C%20Stride%20Toward%20Freedom. Rodell was also able to sell King's book to be made into a film. https://www.universityarchives.com/auction-lot/martin-luther-king-jr.-tls-re-stride-toward-fr_6549cd933b.

154. Conservation Foundation, *Implications of the Rising Carbon Dioxide Content of the Atmosphere*, 1963, https://babel.hathitrust.org/cgi/pt?id=mdp.39015004619030&view=1up&seq=1. Rachel Carson was friends with Revelle and regularly read Conservation Foundation Reports. Also see, Bob Musil, "The Tragedy of Climate Change Reports," *Huffington Post*, January 20, 2015, https://www.huffpost.com/entry/the-tragedy-of-climate-ch_b_6502804?guccounter=1&guce_referrer=aHR0cHM6Ly93d3cuZ29vZ2xlLmNvbS8&guce_referrer_sig=AQAAAIj48n8HWxB5lITIJ5irh13-qltf4WifOJ63J16k2bDKXEv2N2t-RMxVGs655MZHraWPkT6kcq_H_4MUN0OwN2j4-7Rwe0-bYIbYc0kw5zT_ZmJKAcaMBV0lwUomCxvo7Rf-P1hTFfntM13mlnjhSNQMreUmtAEMtdX_Zi9klbU1D.

155. See *Blast Zone: Natural Gas and the Atlantic Coast Pipeline: Causes, Consequences and Civic Action* (Rachel Carson Council, 2018) at https://rachelcarsoncouncil.org/wp-content/uploads/2017/09/blast-zone-final.pdf.

156. Claudia Steiner, "Injustice Along the Cape Fear River" at https://rachelcarsoncouncil.org/environmental-injustice-along-cape-fear-river/ and "Cape Fear Cruise Exposes Environmental Injustice" at https://rachelcarsoncouncil.org/cape-fear-cruise-exposes-environmental-injuRACstice/.

157. See "RCC Showcases National Wood Pellet Forum" at https://rachelcarsoncouncil.org/rcc-showcases-national-wood-pellet-forum/.

158. For the RCC video documentary *Wood Pellet Wasteland* and other short

videos, see https://rachelcarsoncouncil.org/industrial-wood-pellets-false-climate-solution/. Also, among others, see the Rachel Carson Council's full page ad on the back cover of T*he New Republic*, January/February, 2026.

159. See Robert K. Musil, "Rachel Carson, Cancer and the NIH" at https://rachelcarsoncouncil.org/rachel-carson-cancer-and-the-nih/ and various essays in his "Connections" blog opposing Donald Trump and calling for his impeachment starting in his first term with Musil, "The Environment and Impeachment?" at https://rachelcarsoncouncil.org/the-environment-and-impeachment/ and calls for impeachment throughout his two terms with Musil, "The Constitution and Donald Trump," at https://rachelcarsoncouncil.org/the-constitution-and-donald-trump/ and Musil, "Return to Normalcy?" at https://rachelcarsoncouncil.org/return-to-normalcy/.

160. https://rachelcarsoncouncil.org/voter-toolkit/.

161. See https://beinecke.library.yale.edu/exhibitions-visiting/upcoming-exhibitions.

162. The National Conservation Training Center (NCTC) in Shepherdstown, West Virginia, https://www.fws.gov/program/national-conservation-training-center houses Rachel Carson materials and artifacts belonging to the USFWS, as well as the archives of the Rachel Carson Council. Mark Madison, Ph.D., an environmental historian, is the head of the NCTC's Museum and Archives and a member of the Rachel Carson Council National Advisory Council. https://rachelcarsoncouncil.org/about-rcc/rachel-carson-council-rcc-national-advisory-council/.

163. For the Eden Hall Campus, the Falk School of Sustainability, and various Rachel Carson Awards, see https://www.chatham.edu/academics/colleges-and-schools/falk-school-of-sustainability-environment/index.html. Similarly, Duke University awards Rachel Carson Scholars in Marine Science.

164. The Springsong Museum is holding numerous activities and programs even before the museum itself is completed. https://www.springsongmuseum.org/.

165. See The National Portrait Gallery on Hanbury's bust of Carson at https://npg.si.edu/learn/classroom-resource/rachel-carson-1907%E2%80%931964.

166. The Rachel Carson Memorial Statue by David Lewis was completed in 2013 on the waterfront of Woods Hole Cove. https://www.atlasobscura.com/places/rachel-carson-memorial. That Carson's iconic status continues is evident in yet another bronze statue of her begun by artist Paula Slater

in 2025 as part of a competition to create a bronze bust to be placed at the Department of the Interior building where Carson worked. https://paulaslater.com/sculpture/rachel-carson-portrait-bust-clay-for-bronze/.

167. Lear, *Witness*, 180.

168. Lear, *Witness*, 477. The Sierra Club's Rachel Carson Society is at https://myplan.sierraclub.org/rachel-carson-society.

169. Douglas Brinkley, *Silent Spring Revolution: John F. Kennedy, Rachel Carson, Lyndon Johnson, Richard Nixon, and the Great Environmental Awakening* (New York: Harper, 2011), 369. For Carson and the Defenders of Wildlife, see https://defenders.org/blog/2022/03/women-have-long-led-defenders-pack.

170. For Jamie Rapaport Clark and the influence of Rachel Carson on her life and career, see https://defenders.org/blog/2021/08/open-letter-future-women-conservation.

171. See The Nature Conservancy's recognition of Rachel Carson's work at https://www.nature.org/en-us/about-us/where-we-work/united-states/maine/stories-in-maine/the-nature-conservancy-in-maine-rachel-carson-walking-in-her-footsteps/?en_txn1=p_gr.gd.eg.x.storiesandnews.dynamic.gh&gad_source=1&gad_campaignid=23483423232&gclid=Cj0KCQiAtfXMBhDzARIsAJ0jp3AhO9cBn6MkyffUavorBoJKODq4AUGHy3oiuR-uYWb0kt_YmJHKJ3MaAhHxEALw_wcB. Also see Lear, *Witness*, 477, for Carson's bequest to The Nature Conservancy. For the Rachel Carson National Wildlife Refuge, see https://www.fws.gov/refuge/rachel-carson.

172. Lear, *Witness*, 408, 418.

173. For Carson's response to her critics published by Audubon, see https://www.audubon.org/magazine/archives-rachel-carson-answers-her-critics.

174. For Audubon's Rachel Carson Award, see https://www.audubon.org/about/rachel-carson-award?int_src=site_search&int_query=Rachel%20Carson%20Award. For Rachel Carson Award recipients, see https://en.wikipedia.org/wiki/Rachel_Carson_Award.

175. For Carson's writing for *St. Nicholas Magazine*, see Lear, *Witness*, 18-20. For *Holiday*, see Lear, *Witness*, 127-128 and for the *Woman's Home Companion* see Lear, 283-285. For Carson and the *Baltimore Sun*, see Lear, *Witness*, 79, 83.

176. For the Society of Environmentalists (SEJ) Rachel Carson Book Award,

see https://www.sej.org/rachel-carson-environment-book-award-sej-23rd-annual-awards-reporting-environment#:~:text=The%20Society%20of%20Environmental%20Journalists%20(SEJ)%20awards,Think%20About%20Animals*%20by%20Christopher%20J.%20Preston. For recent winners, including Tatianna Schlossberg, see https://rachelcarsoncouncil.org/voter-toolkit/ and SEJ's memorial for Schlossberg at https://www.sej.org/sej-mourns-loss-member-Tatiana-Schlossberg .

177. Lear, *Witness*, 400, 404-406.

178. For the Rachel's Network Catalyst Award, see https://rachelsnetwork.org/catalystawardees25/.

179. Lear, *Witness*, 181-182.

180. For Carson's predecessor, Florence Merriam Bailey, see Musil, *Rachel Carson and Her Sisters*, 34-41.

181. Lear, *Witness*, 181.

182. Robert K. Musil, *Washington in Spring: A Nature Journal for a Changing Capital* (Bartleby Press/Rachel Carson Council, 2016.) Available through the Rachel Carson Council at https://rachelcarsoncouncil.org/product/washington-in-spring/ and in a Kindle edition through Amazon at https://www.amazon.com/Washington-Spring-Journal-Changing-Capital/dp/0935437460 . Also see Musil's nature essays in the footsteps of Rachel Carson along the C&O Canal such as "Seeing Through the Eyes of Rachel Carson" https://rachelcarsoncouncil.org/seeing-eyes-rachel-carson/, "Can Springtime Soothe Us?" https://rachelcarsoncouncil.org/can-springtime-soothe-us/, "I'm Nobody, Who Are You?" https://rachelcarsoncouncil.org/im-nobody-who-are-you/, and others in his blog, "Connections." https://rachelcarsoncouncil.org/category/blog/.

183. The full text of Rachel Carson's June 12, 1962 commencement address at Scripps College for Women is on Iowa State University's Catt Center website at https://awpc.cattcenter.iastate.edu/2025/07/16/of-man-and-the-stream-of-time-june-12-1962/.

Photos and Illustrations

Cover. Rachel Carson in dark coat and bow at about age 12. Rachel Carson Council.

Page 1. Rachel Carson reading to her dog Candy. Rachel Carson Council.

Page 2. *Under the Sea-Wind* book cover. Raptis Rare Books.

Page 3. (left) American eel (Anguilla). Wikimedia Commons. Brevoort, James Carson, 1818-1887. (right) Mackerel (Scomber). Sherman F. Denton, *Game Fish of New York*. circa. 1895-1907.

Page 4. *Silent Spring*, 1962 first edition book cover. baywindowbooks.

Page 5. Rachel Carson on bench with Grace Croft at Pennsylvania College for Women, Rachel Carson Council.

Page 6. Biological Laboratory of the U.S. Bureau of Fisheries, Beaufort, North Carolina. Freshwater and Marine Image Bank, University of Washington.

Page 7. Hendrik Van Loon letter to Rachel Carson. Rachel Carson Papers, Yale Beinecke Rare Book and Manuscript Library.

Page 8. Black Skimmer feeding. Wikimedia Commons. Photo: Charles J. Sharp https://creativecommons.org/licenses/by-sa/2.0/deed.en.

Page 10. *Silent Spring* illustration to Chapter 7, "Needless Havoc" by Lois and Louise Darling. Rachel Carson Council.

Page 11. *H.M.S. Challenger*. Wikipedia Commons.

Page 12. William Beebe at porthole, *National Geographic Magazine*, June, 1931. © William Beebe/WCS. Reproduced by permission of the WCS Archives.

Page 12. Bostelmann painting for Beebe, *National Geographic Magazine*, June 1931. © Wildlife Conservation Society. Reproduced by permission of the WCS Archives.

Page 13. Copepod. Image Quest Marine, photo by Peter Parks.

Page 14. Tetraphalia (Ctenophore). Woods Hole Oceanographic Institute, photo by Larry Madin.

Page 15. Rachel Carson on dock at Woods Hole, MA. Rachel Carson Council, photo by Edwin Gray.

Page 17. Black skimmer, illustration for 1991 edition of *Under the Sea-Wind*. Bob Hines. Courtesy of USFWS Museum/Archives.

Page 19. The Wreck of the Mary B., illustration by Bob Hines. Courtesy of the USFWS Museum/Archives.

Page 20. Anglerfish, photo by David Shales/Minden Pictures.

Page 21. American eel. Courtesy of Robert S. Nicholson.

Page 23. Great Blue Heron, illustration by Bob Hines. Courtesy of the USFWS Museum/Archives.

Page 23. Hermit Crab, illustration by Bob Hines. Courtesy of the USFWS Museum/Archives.

Page 24. Ghost Crab, illustration by Bob Hines. Courtesy of the USFWS Museum/Archives.

Page 25. Sanderling on nest in Arctic. Photo by Jereon Reneerkens.

Page 26. Purple Saxifrage in Arctic spring, Wikimedia Commons. Photo: Alastair Rae. https://creativecommons.org/licenses/by-sa/2.0/deed.en.

Page 27. Anchovies, illustration by Bob Hines. Courtesy of the USFWS Museum/Archives.

Page 28. Mackerel, photo courtesy of Robert S. Nicholson.

Page 29. Glass eel, photo courtesy of Robert S. Nicholson.

Page 31. Mullets jumping in migration, *Hakai Magazine*.

Page 33. Henry Beston cottage on Cape Cod, The Henry Beston Society.

Page 35. Chandler Robbins, All About Birds, photo by Barbara Dowell/USGS.

Page 35. *Birds of North America, A Golden Book Guide*, book cover, Thiftbooks.

Page 36. Paul Brooks, photo PBS Channel 21, Rochester, NY. Earth Week, 1972.

Page 37. *Salar the Salmon* book cover, The Cary Collection.

Page 38. Old Berry Hall I, Chatham University Archives and Special Collections.

Page 40. Snowy Owl, illustration by Bob Hines. Courtesy of the USFWS Museum/Archives.

Page 41. Anna Botsford Comstock, *Entomology Today.*

Page 41. Florence Merriam Bailey, Wikipedia.

Page 42. *St. Nicholas Magazine* cover with Rachel Carson prize at 14 for "My Favorite Recreation," Harropian Books.

Page 43. Rachel Carson on Hawk Mountain, PA. Rachel Carson Council. Photo by Shirley A. Briggs.

Page 44. Eel drawings, *Fishes of the Gulf of Maine.*

Page 45. Mullet drawings, *Depths of the Ocean.*

Page 46. Rachel Carson and Eric Sevareid in her home study, CBS News, "The Silent Spring of Rachel Carson," April 3, 1963, photo from CBS Photo Archive/ Getty Images.

Page 47. Linda Lear, *Rachel Carson: Witness for Nature*, cover photo, Amazon.com.

Page 49. Rachel Carson exploring tidal pool near her Maine cottage, Rachel Carson Council, photo by Shirley A. Briggs.

Page 51. Prochlorococcus, Wikipedia Commons.

Page 52. House collapse in ocean, Buxton, North Carolina, Associated Press.

Page 53. Trash on beach, Great Lakes Now.

Page 55. John F. Kennedy and Stuart L Udall, Stewart L. Udall papers, University of Ariona, Tucson, Special Collections.

Page 57. Evening at Woodend Mansion, The Knot/Woodend Sanctuary and Mansion.

Page 58. Rachel Carson writing in her study, Rachel Carson Council.

Page 59. Rachel Carson Council President & CEO, Robert K. Musil, Rachel Carson Council.

Page 60. Sen. Sheldon Whitehouse speaking on climate, Senate Television.

Page 60. Sen. Chris Van Hollen at RCC Jubilee, Rachel Carson Council.

Page 62. Elsie Herring, Environmental Health News, photo by Alex Boerner.

Page 62. Drowned Pigs after Hurricane Florence, photo by Rick Dove.

Page 63. Martin Luther King and *Stride Toward Freedom* cover, Book Marks.

Page 64. ACC pipeline protest, *Blast Zone*, Rachel Carson Council, photo by Chesapeake Climate Action Network.

Page 65. RCC Cape Fear Riverboat Cruise at Riverwalk, Wilmington, North Carolina, Rachel Carson Council.

Page 67. Demonstration against science cuts at NIH. Rachel Carson Council, photo by Bob Musil.

Page 69. Rachel Carson bronze statue at Woods Hole by David Lewis, Wikipedia.

Page 70. Mary Hazell Harris, Defenders of Wildlife, photo by Karl Umbriaco.

Page 71. Jamie Rapaport Clark, Wikipedia.

Page 72. Dorceta Taylor, Rachel Carson Award winner, Yale School of the Environment.

Page 73. Tatiana Schlossberg, JFK Library Foundation.

Page 74. LaTricea Adams, Rachel's Network 2025 Catalyst Award winner, Rachel's Network.

Page 76. C&O Canal with tavern and barge, Wikipedia.

Page 77. Rachel Carson entrance photo for Johns Hopkins Graduate School. Rachel Carson Council.

Page 78. Rachel Carson Commencement Address, Scripps College for Women. Claremont Archives.

Page 79. RCC student activists at Capitol, RCC Advocacy Days. Rachel Carson Council.

Acknowledgements

Becoming Rachel Carson began life during the COVID-19 epidemic as a lengthy introduction to a new scholarly and sumptuous edition — replete with updated marine science, an annotated text, and a trove of contemporary and historical photographs and illustrations — of Rachel Carson's first book, *Under the Sea-Wind*. I had been encouraged to produce it by my editor at the Rutgers University Press, Peter Mickulas, a fine environmental historian, who had shepherded me through my first book about Rachel Carson, *Rachel Carson and Her Sisters: Extraordinary Women Who Have Shaped America's Environment.*

As I worked on the manuscript, I was at the same time increasingly consumed by my passion to build and expand the Rachel Carson Council (RCC) which grew rapidly in size, staff and influence, The RCC happily engaged more and more of my time and energy, finally devouring what little was left to complete what proved to be a far more challenging project than I had ever imagined. As a result, as the years passed, and given changes in the budgets and markets for books from academic presses, Rutgers Press sadly decided that it could no longer afford to publish the Robert K. Musil edition of *Under the Sea-Wind.* But, happily, *Becoming Rachel Carson* was born. It is, in effect, a rewriting and expansion of that overdue Introduction, with some of the best photos and illustrations that would have accompanied it, creating this small book, laden with pictures, that traces how Rachel Carson came not only to write *Under the Sea-Wind*, but to become an American writer, scientist, and advocate whose influence is both vast and enduring.

And so, I remain deeply grateful to Peter Mickulas of Rutgers Press whose encouragement over the years has led to my becoming a kind of Rachel Carson authority. I also am grateful to Linda Lear, *the* Carson authority, whose prodigious work has done so much to put Rachel at the center of American life and literature. She also encouraged me to create a new edition of *Under the Sea-Wind* that would have built upon her own classic centennial edition of *Under the Sea-Wind* published by Penguin Press in 2007.

Of course, I could not have done all that is involved in researching an edition of *Under the Sea-Wind* without the brilliant assistance of Rachel Carson Council Fellows who worked directly for me and also became Carson devotees. Kendall Jefferys, Audrey Magnuson, and Emma Brentjens, all of Duke University, carried out the lengthy process of updating the current state of marine science that Rachel Carson used in the 1930s. Joy Reeves, also of Duke, and now a senior leader at the Rachel Carson Council as Director of Policy and Strategic Development, brought me right up to the finish line. I am indebted to each of them.

The Rachel Carson Council Board of Directors has consistently encouraged me to write and publish while running the RCC and given me the leeway to follow my Carson mania. I am especially grateful to our Chair, Roger Christie, Rachel Carson's son, who holds the rights to his mother's work and who has made simple the business of literary rights. And, to this day, I am glad that when interviewing with the Board to become the CEO of the RCC, Roger welcomed my quite political approach to his mother's legacy and planted a seed by asking if I would at some point involve the Rachel Carson Council in caring about the ocean. The RCC's Coasts and Ocean Program and, years later, *Becoming Rachel Carson*, are the result.

I have also been inspired every day by the Rachel Carson Council staff, especially by Joy Reeves and Diego Tovar, Director of Campus and Civic Engagement, who are published authors themselves. Diego has already published eight environmental sci-fi novels and has been instrumental in helping me and the Rachel Carson Council in finding our way along the winding path to publishing a book without the use of a publishing house. And all of our work has been made far more visible and accessible, on line and in the media, with our followers much increased by Sydney O'Shaughnessy, RCC's Communications Director, who has brought us into the modern age and already begun marketing *Becoming Rachel Carson*.

I cannot say enough about my colleague and friend, Mark Madison, the historian and archivist of the U.S. Fish and Wildlife Service and its conference center, museum, and archives at the National Conservation Training Center (NCTC) in Shepherdstown, WVA. Mark is one of the most generous, selfless, and friendly people I have ever known. Endowed with a sunny disposition and an infectious laugh, I first met Mark at a book launch event for *Rachel Carson and Her Sisters* at the American Society of Environmental Historians (ASEH) where he introduced me to everyone there as if I were Rachel Carson herself. Since then, he has had me speak on Rachel Carson at the NCTC, read my manuscripts, appeared with me at numerous events about Rachel Carson, and, importantly, helped me pack up, load and then drive back to West Virginia, his van full of RCC archival material, including Rachel's personal library, from the Rachel Carson Landmark Alliance home in Silver Spring. Mark also made available from the NCTC archives the complete set of original illustrations that Rachel's friend and USFWS colleague, Bob Hines, made for a 1991 edition of *Under the Sea-Wind.* They were to be used for my Rutgers Press edition, but now some of the most interesting ones grace the pages of *Becoming Rachel Carson.*

Becoming Rachel would not exist at all without the tireless work of Ross Feldner, the web and design consultant of the Rachel Carson Council, who laid out and designed my first RCC book (with Bartleby Press), *Washington in Spring: A Nature Journal for a Changing Capitol.*

Over the years since I started at RCC when Ross had been the sole remaining staff keeping the lights on, maintaining records, and updating our web site, I have worked closely with him every day as he lays out and posts all of RCC's many newsletters, essays by our Fellows, daily and weekly environmental news, action alerts, appeals, and more. Ross completely designed our current web site and has become RCC's most popular and followed writer with his feature "The Bird of the Week" and original essays on the life and lore of birds. On top of all this, Ross suggested that we could publish *Becoming Rachel Carson* on our own. He has designed the book, enhanced its photos and illustrations, and entered endless edits and revisions in drafts too numerous to mention. But I am grateful not just for Ross' unlimited labor that holds the RCC together and has made this book possible. It is Ross' old-fashioned work ethic, his positive, can-do approach to mountains of material, and his humor, his political knowledge and commitment that makes each of our conversations a delight. He keeps me going. Thank you.

No acknowledgements would be complete without my wife, Caryn McTighe Musil, one of the nation's leaders in higher education, democratic rights, feminism and justice, to whom I dedicated *Rachel Carson and Her Sisters*. She has been at my side offering advice, professional insights, encouragement, and love since we first met in the days of "miniskirts and mass mobilizations." It is Caryn who urged me, or, rather, insisted, that I find a way to finish my Rutgers edition of *Under the Sea-Wind*. But, if not, to share with the public somehow all that I have learned and admired about Rachel Carson, the environmental movement and our nation's future environmental leaders to which she gave birth. Caryn, thank you.

About the Author

Becoming Rachel Carson: The Origin and Legacy of a Legend is environmental leader and advocate Bob Musil's fourth book on the American environment and the culmination of his longstanding respect for and fascination with Rachel Carson. Musil is the President and CEO of the Rachel Carson Council, the legacy organization envisioned by Rachel Carson and founded in 1965 after her death from breast cancer. Prior to rebuilding the RCC in 2014, Musil was the longest-serving CEO of the Nobel Peace Prize-winning Physicians for Social Responsibility and initiated their global climate change and environmental health programs. Musil is also the author of *Hope for a Heated Planet: How Americans Are Fighting Global Warming and Building a Better Future* (Rutgers); *Washington in Spring: A Nature Journal for a Changing Capital* (RCC/Bartleby); and the 2016 Choice outstanding book, *Rachel Carson and Her Sisters: Extraordinary Women Who Have Shaped America's Environment* (Rutgers).

Trained in environmental health, literature and the humanities, Musil holds degrees from Yale, Northwestern, and the Johns Hopkins School of Public Health and honorary degrees in science and humane letters. Bob has taught environmental politics and history at American University and politics and American religion at the Wesley Theological Seminary in Washington, D.C. as well as American Studies, English literature, history and politics at Temple, Northwestern, and La Salle Universities.

Praise for Rachel Carson and Her Sisters

"This is a long overdue book, giving great credit to the long line of women who have done so much to shape our culture's view of the world around us and of our prospects in it. We desperately need that culture to heed their words!"

—Bill McKibben, author *Oil and Honey: The Education of an Unlikely Activist*

"Rachel Carson is only the best-known example amidst an inspiring cast of pioneering and modern women environmental leaders that Musil brings to life. Readable, reliable, and rousing—a book for anyone who cares about America's past and future."

—Gene Karpinski, President and CEO of the League of Conservation Voters

"A treasure! A welcome discovery of the linkages among profoundly caring, ecologically-aware women across time, and the truths of our ecological crisis. Musil shows clear-eyed science and heartfelt story telling are not mutually exclusive."

—Rebecca Wodder, former President and CEO of American Rivers

"Bob Musil provides an important contribution to the history of the environmental movement. He paints a compelling portrait of Rachel Carson and the remarkable women who preceded her and who continue her legacy. He reminds us of the struggles and achievements of Ms. Carson and, just as significantly, the pivotal and courageous role that women have played in fighting for a safer and healthier world."

—Tom Udall, US Senator, New Mexico

"With deep grounding in women's history, environmentalism, and public health—and, just as importantly, with great reverence—Musil introduces us to a pantheon of remarkable women, true heroines everyone. This book offers a new perspective, countless wonderful stories, and inspiration. A great read!"

—Howard Frumkin, Dean, University of Washington School of Public Health

"An absolutely wonderful book! Bob Musil shows Rachel Carson not as a lone voice, but an eloquent one who drew inspiration from-female predecessors and those around her. He argues persuasively that we can understand Carson better if we see her in relation to other women, to the broader environmental movement, and to working in community. Should be required reading for anyone interested in where we have been, and where we need to go."

—Geoffrey Chase, author of *Sustainability in Higher Education*

"An eloquent and moving tribute to the women at the heart and soul of the environmental movement. It is a story of brilliant science, courage, stamina, and a passion for life. We are in debt beyond counting to them and to Robert Musil for telling their stories so well."

—David W. Orr, Paul Sears Distinguished Professor of Environmental Studies and Politics, Oberlin College

"A richly detailed, documented, and eloquent history—a groundbreaking account of undaunted American women, determined to prevent environmental catastrophe."

—Lawrence Wittner, *Huffington Post*

"A great read for anybody who is interested in learning about Rachel Carson's role in a delicate web of connections that makes up the environmental movement. Also, if anybody is interested in the human aspect, the personal lives, and the trials of each of these women, this book certainly will deliver. Musil has stitched together a wonderful collection of true stories about the amazing women who have changed, and are continuing to change, the way we see the world."

—*The Prairie Naturalist*

"Bob Musil brilliantly documents the rich trajectory of women's intellectual and political influence, not just on environmentalism but on public policy and activism. Musil offers fascinating details of Rachel Carson's struggles to be taken seriously as a scientist and unearths the stories of the women—unsung heroes all—who influenced her. A must read for anyone interested in American history, science and environmental politics."

—Heather White, executive director, the Environmental Working Group

"A vibrant, engaging account of the women who preceded and followed Rachel Carson's efforts to promote environmental and human health. In exquisite detail,Musil narrates the brilliant career and efforts of pioneering women from the 1850s onward to preserve nature and maintain a healthy environment. Anyone interested in women naturalists, activists, and feminist environmental history will welcome this compelling, beautifully-written book."

—Carolyn Merchant, author of *The Death of Nature,* and professor of environmental history, philosophy, and ethics, University of California, Berkeley

"Musil contextualizes *Silent Spring* as the culmination of decades of work by other women in science, who were consistently overlooked, underappreciated, and dismissed by their male peers and institutions. These ladies ranged from Victorian garden observers to die-hard chemists and marine biologists. They are tied together by a fierce sense of activism and beautiful writing. Their writing is what drew Musil in. He too wants 'people to connect with science in an approachable way.'"

—Sierra Club Greenlife

"In *Rachel Carson and Her Sisters*, Musil fills the gap by placing Carson's achievements in a wider context, weaving connections from the past through the present. Readers will find new insight into Carson and contemporary figures she influenced... who have historically received less attention. Musil's respect and enthusiasm for these women is evident throughout the book, making it a deeply engaging and enjoyable read. A valuable addition to scholarship on Rachel Carson, female environmentalists, and the American environmental movement in general. Highly recommended. All academic and general readers."

—*Choice*

"Musil uses the life and writings of Rachel Carson as an exemplar of women's participation in the American environmental movement. He places Carson's achievements in contexts by illuminating the lives of trailblazing female scientists who inspired her and for whom she, in turn, paved the way. Extremely well-researched."

—*Foreword*

"*Rachel Carson and Her Sisters* makes a number of important contributions to both environmental history and women's history. Musil's genius is weaving his intriguing, thoroughly researched mini-biographies of individual women into a cohesive larger story of overlapping and mutually reinforcing actions and ideas."

—Environmental History

"In celebrating Rachel Carson's work, Musil takes on the important task of contextualizing this environmental luminary within a tradition of women's research, activism, and authorship."

—Women's Review of Books

"This book is one-of-a-kind. Musil provides a remarkable new perspective on the role of individual women in the US environmental movement."

—Cathy Middlecamp, Professor, Nelson Institute of Environmental Studies, University of Wisconsin-Madison

Praise for Becoming Rachel Carson

"Among its many other virtues, this slim volume captures Carson's grace and eloquence as a writer—a translator of science into prose that all could understand. That's a great gift to have given the world."

—Bill McKibben, author of *The End of Nature*

"In *Becoming Rachel Carson*, Bob Musil has done a wonderful service to one of the pioneers of the environmental movement. This account of Rachel's life and work illustrates just how far ahead of her time she was, with a legacy that continues to bloom."

—U.S. Senator Sheldon Whitehouse, author of *Captured: The Corporate Infiltration of American Democracy*

"Rachel Carson's voice is timeless. Her writing is a luminous gift to the world. At a time when humanity is moving ever farther from its essential connection to nature, we've never needed her more. *Becoming Rachel Carson: The Origin and Legacy of a Legend* will delight everyone who revels in Carson's work—and I count myself among them. Beautifully illustrated and written straight from the heart, this book by Bob Musil is yet another gift to us all."

—Susan Casey, *New York Times* best-selling author of *The Underworld: Journeys to the Depths of the Ocean*

"What more is there to learn about Rachel Carson? A lot, it turns out, and Bob Musil is our excellent teacher in *Becoming Rachel Carson*. In clear and passionate prose, Musil shows Carson's evolution from sensitive farm girl to brilliant biologist to ocean-obsessed best-selling author — all of which led logically to her masterpiece *Silent Spring*. Along the way, who knew that Carson was well aware of — and wrote about — the dangers of climate change as far back as the

1950s? With his own deep experience as an environmental leader and author, Musil deftly lays out how Carson's pioneering insights and truly lyrical prose made her the inspiring environmental leader we all want to be."

—Mike Tidwell, author, *The Lost Trees of Willow Avenue: A Story of Climate and Hope on One American Street*; founder and Executive Director, Chesapeake Climate Action Network

"For those who admire Rachel Carson for her courageous environmental advocacy, Bob Musil deftly draws them to her earlier writing that exhibits profound sensitivity to earth's smallest creatures, a sensitivity that infuses and motivated all her work. As he demonstrates in his wonderful new book, *Becoming Rachel Carson: The Origin and Legacy of a Legend,* and in all of his own work for the Rachel Carson Council and beyond, Musil is also propelled by a deep love for earth's ecosystems and all its creatures."

—Melanie Choukas-Bradley, author of *City of Trees, A Year in Rock Creek Park, Finding Solace at Theodore Roosevelt Island* and other nature books

"If ever we needed a role model for confronting the immense ecological and climate challenges facing our planet, Bob Musil reminds us why Rachel Carson remains one. In this compelling new biography, Musil distills the lessons of Carson's extraordinary life into a work that should be required reading for anyone seeking inspiration, courage, and hope."

Carson combined scientific knowledge, imagination, empathy, and moral conviction to create powerful narratives that revealed, through vivid imagery and emotional experience, the inextricable connections between people and the natural world. Musil shows that *Silent Spring* was not the beginning of Carson's influence but the culmination of a lifetime devoted to science, observation, writing, and the painstaking work of building credibility. The result was a book that helped catalyze a movement.

At a moment when global challenges can seem to be accelerating beyond our control, Musil offers timely lessons about how one person can make a profound difference—and why Carson's example matters now more than ever."

—Toddi Steelman, Vice President/Vice Provost for Climate and Sustainability; Professor of Environmental and Natural Resource Policy, Duke University

"Bob Musil's beautiful slim book, *Becoming Rachel Carson*, traces the sources and story of the rise to fame of the 20th century's most influential environmental writer. Carson's *Silent Spring* (1962) kick started the modern environmental movement, but Musil traces Carson's importance more than two decades earlier. Through a careful examination of Carson's early life and career and her undeservedly forgotten first book, *Under the Sea-Wind* (1941), Musil uncovers Carson's literary and scientific genius for the remainder of her life. *Becoming Rachel Carson*, replete with illustrations and photographs, then recounts the amazing array of individuals and organizations who still carry out Carson's legacy. Perhaps the best short introduction to Rachel Carson and her influence."

—Mark Madison, Historian, U.S. Fish and Wildlife Service; Heritage, Library, and Partnerships Manager, National Conservation Training Center.

www.ingramcontent.com/pod-product-compliance
Lightning Source LLC
LaVergne TN
LVHW051129210926
848482LV00001B/26
* 9 7 9 8 2 3 4 0 8 0 6 7 7 *